KV-681-709

# Laboratory Exercises
# in Zoology

# Laboratory Exercises in Zoology

H. S. Luker, B.Sc., M.I.Biol.
and
A. J. Luker, L.C.P., M.I.Biol.
*Head of the Biology Department, The Lakes School, Windermere*

**LONDON**
**BUTTERWORTHS**

THE BUTTERWORTH GROUP

England: Butterworth & Co. (Publishers) Ltd.
London: 88 Kingsway, WC2B 6AB
Australia: Butterworth & Co. (Australia) Ltd.
Sydney: 20 Loftus Street
Melbourne: 343 Little Collins Street
Brisbane: 240 Queen Street
Canada: Butterworth & Co. (Canada) Ltd.
Toronto: 14 Curity Avenue, 374
New Zealand: Butterworth & Co. (New Zealand) Ltd.
Wellington: 49/51 Ballance Street
Auckland: 35 High Street
South Africa: Butterworth & Co. (South Africa) (Pty) Ltd.
Durban: 33/35 Beach Grove

590 72
July 2nd ✓

First published 1971

© Butterworth & Co (Publishers) Ltd. 1971

Suggested U.D.C. number: 591·1(076·5)
Suggested Additional number: 591·08
ISBN 0 408 57850 5

CHRIST'S COLLEGE
LIBRARY
LINE STACK
Access on No. 40413
Class No. 590·72·
Catal. 19.2.74 CE

CC101025

Filmset by Filmtype Services, Scarborough, Yorkshire
Printed by Chapel River Press, Andover, Hampshire

# Contents

# Preface

These exercises are most suitable for sixth form students studying for Advanced level Zoology or Biology. However, a large proportion of the practical work suggested here could be performed quite successfully by pupils at a lower level in school.

The exercises are concerned mainly with physiology and although some dissection techniques are described this aspect of investigation has been neglected largely because we feel that there are school texts which cover quite adequately dissection methods.

This book has been compiled to provide a compact collection of information required by teachers, technicians and students in practical classes. In selecting the practical work for inclusion we have taken into consideration the demands made by the Zoology and Biology syllabuses of the various Examination Boards. The modern approach to the teaching of biology stresses the importance of practical work. As many schools are without a full time biology laboratory technician, it is often necessary for teachers and pupils to look after livestock and set up apparatus for laboratory investigations. We have taken care over choosing the type of animal whose physiology is to be examined and we have named only those animals which may be accommodated successfully in a school laboratory and which may be obtained from one of the biological suppliers. The materials required for each exercise are listed in each case and where applicable, the names and addresses of suppliers have been included. We have tried to use equipment which is not too costly and where it is possible, we have given details of the construction of some of the pieces of apparatus.

To make the exercises more meaningful we have included several short theoretical introductions but out of necessity these have been kept to a minimum.

We hope that the results of some of these investigations will stimulate further enquiry.

It is impossible to acknowledge the source of all the material included here particularly as some of the exercises have now become traditional in an introductory course in zoology. Nevertheless, we should like to thank the many unnamed colleagues, friends and pupils who have provided us, sometimes unwittingly, with ideas for this text.

Windermere                                                    H.S.L.
1969                                                             A.J.L.

# Acknowledgements

We should like to express our thanks to Mrs. E. M. Ramsden who typed the manuscript and Mr. Stuart Thompson who took the photographs for Plates D2, D3, D4, D5, E1 and E2.

We are most grateful to Dr. M. E. Wallace of the Department of Genetics, Cambridge University for her help and advice on the use of mice for teaching genetics. We should also like to thank Dr. Bruce Hobson of the Royal Infirmary, Edinburgh for advice on the care of Xenopus and Mr. J. Haller of Harris Biological Supplies for detailed information on some of the stock animals discussed in the text.

We are indebted to the following who have given permission to quote from their publications or work: The Anti-Locust Research Centre; the Editor of The School Science Review; the Universities Federation for Animal Welfare; the Keeper of the Department of Zoology, The British Museum (Natural History); Blackwell Scientific Publications, Mr. D. Ryman and Mr. M. N. Dunthorne.

We much appreciate the assistance given by the following scientific suppliers: B.D.H. Chemicals, Ltd., T. Gerrard and Co. Ltd., Griffin and George Ltd., Philip Harris Ltd., Kodak Ltd., Matburn Surgical Equipment Ltd., C. F. Palmer Ltd., Shandon Scientific Co. Ltd. and Sandoz Ltd.

# Physical and Chemical Investigations

## 1. DIFFUSION

Diffusion is the movement of molecules or ions from a region of high concentration to a region of relatively low concentration and this continues until they are evenly distributed.

### Diffusion through a membrane

If a 10% solution of glucose is separated from a 20% solution of glucose by a membrane which is permeable to glucose, after a time the solution on each side will contain 15% glucose and a state of equilibrium is reached when there are as many glucose molecules passing one way through the membrane as there are passing in the opposite direction.

Cell membranes are not permeable to all substances in solution, they are selectively permeable.

Provided that the membrane is permeable a substance may diffuse into the cell from the environment without the cell using any energy during the process. But diffusion alone does not account for the way in which all substances enter cells. In some cases cells are able to take in substances which are in low concentration outside the cell compared with a relatively high concentration in the protoplasm. During this process the cell expends energy and materials are said to be obtained by active transport.

1.1.   A SIMPLE DEMONSTRATION OF DIFFUSION

*Materials required*—Test tubes in rack
$\qquad\qquad\qquad$ 150 cm³ beaker
$\qquad\qquad\qquad$ 100 cm³ measuring cylinder
$\qquad\qquad\qquad$ 3·5 g gelatin
$\qquad\qquad\qquad$ Water
$\qquad\qquad\qquad$ 10% copper sulphate
$\qquad\qquad\qquad$ Ammonium hydroxide
$\qquad\qquad\qquad$ Congo red (aqueous solution)

*Procedure*

Dissolve 3·5 g gelatin in 100 cm³ water. Pour some of this solution into 2 test tubes to a depth of about 8 cm. Allow to set.
(1) Add an excess of ammonium hydroxide to a little 10% copper sulphate in a test tube to obtain a deep blue solution. Pour some of this solution on to the jelly in one test tube. Mark the boundary between the two substances and leave the tube to stand for 24 h.
What do you notice about the distribution of the blue solution after this length of time?
(2) To the jelly in the second tube add an aqueous solution of congo red using the same amount as the copper ammonium hydroxide solution. Mark the boundary between the two substances and leave the tube to stand for 24 h.
Do you notice any difference in the result from that obtained in the first demonstration?
Can you suggest a reason for any difference you observe?

FURTHER WORK

(1) Can you suggest why there is an optimum surface area to volume ratio in animal cells?
(2) Is there any difference in the rate of diffusion of oxygen in air and in water?
(3) Try to explain how the rate of diffusion of oxygen in the tracheoles of an insect is increased with an increase in the activity of the insect.
(4) List the features present in the lungs of mammals which favour the diffusion of oxygen and carbon dioxide.

## 2. OSMOSIS

When water is separated from a solution with water as the solvent by a membrane which is permeable to water but not to the solute i.e. a selectively permeable membrane, a greater number of water molecules will pass through the membrane into the solution from the water side than will pass in the opposite direction. Water molecules behave like other molecules and pass from a region of higher concentration to one of lower concentration. The movement of water molecules through a selectively permeable membrane is known as **osmosis**.

Water molecules, like the solute molecules considered in the section on diffusion, are in a constant state of random movement. When a solution is separated from water alone by a selectively permeable membrane water molecules will hit the membrane on both sides and pass through in both directions but for any given area of the membrane more water molecules will hit the membrane

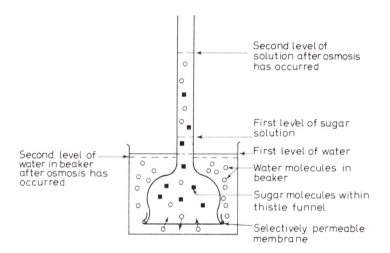

Fig. A1. *Simple apparatus to show the result of osmosis*

on the water side than on the side of the solution as there are less water molecules here and some of the area of the membrane is taken up by bombarding solute molecules which are unable to pass through the membrane.

The results of osmosis may be seen in the traditional demonstration illustrated here (*see Fig. A1*).

*Interpretation of Fig. A1*

The solution of sugar and water will be diluted by the water molecules which pass into the thistle funnel from the beaker. The level of solution in the stem of the funnel will rise as a result. After a time the level of solution remains steady. The pressure of the column of liquid is sufficient to prevent any further passage of water into the solution. This pressure which just prevents osmosis occurring is known as the **osmotic pressure** of the solution.

## Osmotic pressure

The osmotic pressure of a solution is only effective when the solution is separated from the solvent by a selectively permeable membrane. The ability to set up an osmotic pressure (O.P.) is described as the osmotic potential of the solution.

The osmotic pressure of a solution is proportional to the molar concentration of the solute. A solution which contains the gram molecular weight of a non-electrolyte dissolved in 22·4 litres has an osmotic pressure of 1 atmosphere at 0°C and atmospheric pressure.

When two solutions of the same substance in differing concentrations are compared, the one with the greater concentration of solute and thus the higher osmotic pressure is described as a **hypertonic** solution while the other, with the lower osmotic pressure is a **hypotonic** solution. When a hypertonic solution is separated from a hypotonic solution by a selectively permeable membrane osmosis will take place and the hypertonic solution will become diluted.

Solutions with the same osmotic pressure are described as **isotonic.**

Most living cells are surrounded by membranes which are permeable to water so that water may be distributed by osmosis between cells and inter-cellular fluid or between cells and the external environment.

It will be obvious then that when tissues are being examined in the laboratory they should be mounted in a solution which is isotonic with the protoplasm inside the cell membrane. In this way damage to the cells either by shrinkage due to loss of water or by bursting due to a gain of water will be avoided.

**Physiological salt solutions** are solutions of sodium chloride with approximately the same osmotic pressure as the cells being examined. (These salt solutions are sometimes described in biology text books as 'Normal' salt solutions but the term normal here does not imply that the solution contains 1 gram equivalent of NaCl in 1 litre of water.)

A salt solution suitable for examining mammalian cells except blood cells, is prepared by dissolving 9·0 g sodium chloride in 1 litre of water.

A salt solution suitable for examining amphibian tissues, except blood, is prepared by dissolving 6·4 g sodium chloride in 1 litre of water.

## 2.1. DEMONSTRATION OF OSMOSIS

*Materials required*—15 cm length Visking tubing (obtainable in various diameters from Griffin & George, Philip Harris, T. Gerrard & Dutt). With care it may be used many times if kept after first use in water with a little formalin added.

Gas jar

Retort stand and clamp

Length of glass tubing (ext. diam. 5 mm), or length of capillary tubing (see method)

2 rubber bungs, one with hole (for 2·5 cm diameter Visking tubing bung size 23, for 1·875 cm tubing bung size 17)

Saturated sucrose solution

Thin string

Syringe or funnel to enable the filling of the Visking tubing

## *Procedure*

The length of Visking tubing is first soaked in water. The solid bung is inserted in one end (alternatively this end may be knotted using a longer length of tubing) and string tied tightly round the tubing to hold the bung in place. The bung with the hole is inserted in the other end and tied in position.

The tube is then *almost* filled with the sucrose solution using a syringe or funnel. The syringe method is much easier. Insert the length of glass tubing. For a rapid appreciation of the increase in volume in the Visking tubing, capillary tubing may be inserted into the rubber bung in place of the ordinary bore tubing.

Arrange the apparatus as in the diagram, surrounding the tubing with water (*see Fig. A2*).

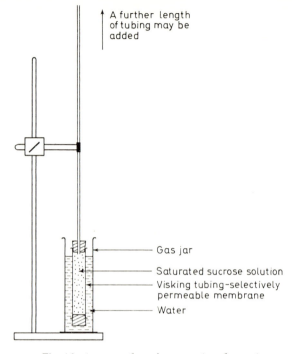

Fig. A2. *Apparatus for a demonstration of osmosis*

## Result

The solution in the tubing will increase in volume and a rise of fluid in the glass tubing should be noted, the speed of ascent and ultimate height of the column being related to the concentration of the sucrose solution.

### FURTHER INVESTIGATION OF OSMOSIS

**Haemolysis**

Haemolysis is the breakdown of the red blood cell membrane resulting in the liberation of haemoglobin. The membrane around a red blood cell functions as a selectively permeable membrane so that when red cells are placed in distilled water or in solutions which are hypotonic to the contents of the cell, osmosis occurs. Water enters the red blood cell which swells as a result and finally bursts liberating the haemoglobin.

2.2. TO DETERMINE THE STRENGTH OF SODIUM CHLORIDE SOLUTION
WHICH CAUSES HAEMOLYSIS

*Materials required*—5 test tubes in a rack

Two, 10 cm³ graduated pipettes or burettes

Microscope

Microscope slides and cover slips

Glass rod

Disposable sterile lancets, obtainable from
Philip Harris or T. Gerrard. If sterile lancets
are not available a clean polished needle
should be flame sterilised just before the
blood sample is required.

Cotton wool

Industrial meths

1 % sodium chloride solution

Distilled water

*Procedure*

Label 5 test tubes A→E. Add 1% NaCl and distilled water in the
following amounts:—

| Tube | 1% NaCl | H₂O | % NaCl |
|------|---------|-----|--------|
| A | 2·0 cm³ | 8·0 cm³ | 0·2 % |
| B | 4·0 cm³ | 6·0 cm³ | 0·4 % |
| C | 6·0 cm³ | 4·0 cm³ | 0·6 % |
| D | 8·0 cm³ | 2·0 cm³ | 0·8 % |
| E | 10·0 cm³ | — | 1 % |

*Blood sampling technique*—The finger from which a sample of blood
is to be taken should be cleaned with industrial meths. The sterile
lancet should then be removed from its packet and used immediately.
When sufficient blood is obtained the lancet should be thrown away.
The finger should then be cleaned again with industrial meths.

*Procedure*

To each of the tubes containing sodium chloride add 3 drops of
blood. Shake each tube and leave to stand for 5 min. Note the

appearance of the contents of each tube. To determine whether haemolysis has taken place you will need to examine microscopically a sample from each tube. Remove a drop of solution with a glass rod from each tube in turn and transfer the drop on to a clean dry slide. You should be able to see why it is important to use a clean and perfectly dry rod for each sample removed. Examine each drop using a high power objective. Note the appearance of any red blood cells.

*Interpretation*

A sample from a tube containing a solution which causes haemolysis will show relatively few, if any, red blood cells when compared with samples from the tubes containing approximately isotonic or slightly hypertonic solutions. At what strength of sodium chloride does haemolysis occur?

As red cells vary in their ability to resist haemolysis in hypotonic solutions it may be interesting to compare your results with those of other members of the class.

2.3. DOES ETHER FUNCTION AS A HAEMOLYSING AGENT?

*Materials required*—1 test tube
                  0·9% sodium chloride solution
                  Ether
                  Disposable sterile lancet
                  Glass rod
                  Cotton wool
                  Industrial meths

*Procedure*

To 5 cm³ of 0·9% sodium chloride solution add a few drops of ether. Shake the test tube well. Obtain a sample of blood as described previously and add 3 drops to the solution in the test tube. Mix well and very gently warm the test tube. Note the appearance of the contents of the tube. Remove a drop and examine as before using a high power objective. Account for the results you observe.

## 3. CHEMICAL INVESTIGATIONS OF SOME ORGANIC COMPOUNDS

### Proteins

Proteins are essential constituents of protoplasm. Every aspect of metabolism is dependent on the catalytic properties of proteins known as enzymes. Proteins have a structural role in the cell and take part in the composition of the cell membrane. Individual proteins have specific roles to play in physiological processes, for example **haemoglobin** is a conjugated protein—a molecule formed by the combination of a simple protein **globin** with a pigment **haem**— and this molecule functions as the respiratory pigment of vertebrates and some invertebrates.

A protein molecule is of colloidal size with a well defined 3 dimensional structure. The great variety in the structure of proteins results from the different arrangements of varying numbers and types of **amino acids**. At present about 30 types of amino acid have been detected as protein constituents and of these around 20 different types occur in the majority of proteins.

The general formula of an amino acid is R.CH.NH$_2$COOH. The difference between amino acids is found in the R. group which may be a single atom e.g. (—H) or a group of atoms e.g. (—CH$_3$). The simplest amino acid is **glycine.**

$$
\begin{array}{c}
NH_2 \\
| \\
H-C-COOH \\
| \\
H
\end{array}
$$

Where R is (—CH$_3$) the amino acid formed is **alanine.**

$$
\begin{array}{c}
H \quad NH_2 \\
| \quad | \\
H-C-C-COOH \\
| \quad | \\
H \quad H
\end{array}
$$

From these formulae it is evident that proteins always contain C,H,O and N. However sulphur is present in many proteins and occasionally phosphorus is found.

An example of an amino acid containing S is **cysteine.**

$$
\begin{array}{c}
H \quad NH_2 \\
| \quad | \\
H-S-C-C-COOH \\
| \quad | \\
H \quad H
\end{array}
$$

The protein, **keratin,** contains a high proportion of cysteine.

Any two amino acids may combine to form a larger molecule known as a **di-peptide.** The amino group ($-NH_2$) of one amino acid condenses with the carboxyl group ($-COOH$) of the other amino acid with the elimination of a molecule of water, an example of a condensation reaction. The bond between the 2 amino acid units is known as a **peptide** bond.

## The formation of a dipeptide from glycine and alanine

The formation of the dipeptide is the first stage in the building of a protein molecule. Each dipeptide has a free ($-NH_2$) and a free ($-COOH$) where additional amino acids may become attached by peptide bonds. When a long chain molecule is formed by the linkage of many amino acids a **polypeptide** results. Proteins are complex polypeptides with molecular weights ranging upwards from several thousands. During the process of digestion, proteins, polypeptides and peptides are **hydrolysed**. The peptide bonds are broken and the constituent amino acids are released.

One of the properties of both amino acids and proteins is their ability to act as **buffers** helping to stabilise the hydrogen ion concentration of a solution. This is important physiologically where the pH of the medium has a marked effect on enzyme catalysed reactions.

In solution the molecules of amino acids and proteins exist as dipolar ions or **zwitterions** (zwitter is the German word for hermaphrodite), they carry both positive and negative charges.

Amino acids and proteins are **amphoteric** substances as they behave as weak acids or weak bases depending on the pH of the solution.

In an acidic solution the amino acid exists largely as a cation. It has an affinity for hydrogen ions, combines with them and thus decreases the hydrogen ion concentration of the solution.

$$R \cdot CH \overset{\displaystyle NH_3^+}{\underset{\displaystyle COOH}{<}}$$

Cation

In an alkaline solution the amino acid exists largely as an anion. It releases hydrogen ions and increases the hydrogen ion concentration of the solution.

$$R \cdot CH \overset{\displaystyle NH_2}{\underset{\displaystyle COO^-}{<}}$$

Anion

The proteins present in blood plasma are able to combine with hydrogen ions thus resisting changes in the pH of the plasma. Haemoglobin inside the red blood cell is negatively charged and acts as an acceptor of hydrogen ions released from the ionisation of carbonic acid formed during the transport of $CO_2$ by the red blood cell.

$$CO_2 + H_2O \longrightarrow H_2CO_3 \overset{\displaystyle HCO_3^-}{\underset{\displaystyle H^+}{<}}$$

$$(Hb)^- + H^+ \longrightarrow (Hb)H$$

Haemoglobin acts as a buffer.

For each protein there is a characteristic **isoelectric point**. This point is the pH value of a solution in which the protein molecule is electrically neutral and there is no tendency of the molecule to migrate towards the anode or cathode in an electric field. The positive charges exactly balance the negative charges. Proteins are most easily precipitated at their isoelectric points.

**Denaturation**

Proteins are susceptible to denaturation. An agent which brings about a change in the spatial structure of a protein molecule is known as a denaturing agent. Denaturation of proteins may be brought about by contact with the ions of heavy metals, strong acids or strong bases, by exposure to heat or by mechanical means. A denatured protein loses its biological properties and those proteins which behave as enzymes lose their catalytic activity after they have been denatured.

## CHEMICAL TESTS FOR PROTEINS

*Materials required for all tests*
> Egg albumen—2% solution
> Gelatin—2% solution. N.B. Gelatin contains
> neither tyrosine nor tryptophan
> Test tubes and test tube holders
> Test tube racks
> Bunsen burners
> Spatulas
> 10 cm³ graduated pipettes

PROTEIN PRECIPITATION

*Reagents*—20% sulphosalicylic acid
> 10% tri-chloroacetic acid (Prevent contact of this reagent
> with the skin)
> Lead acetate solution

*Procedure*

Precipitation by these reagents is accompanied by denaturation of the protein due to the formation of insoluble complexes between the protein and the reagent.

3.1. Treat 5 cm³ protein solution with 2 drops of 20% sulphosalicylic acid at room temperature.

3.2. Treat 5 cm³ protein solution with 1 cm³ of 10% trichloroacetic acid.

3.3 Treat 3 cm³ of an alkaline protein solution with 3 cm³ of lead acetate solution.

The above tests are not specific for proteins alone as these reagents may also be used to precipitate other organic nitrogenous compounds.

## 3.4. REACTIONS RESULTING IN THE FORMATION OF COLOURED COMPOUNDS

### 3.4.1. NINHYDRIN REACTION

*Preparation of ninhydrin solution*—Dissolve 0·2 g ninhydrin in 100 cm³ water. This reagent is unstable and will only keep for 2 days.

*Procedure*

To 3 cm³ of protein solution add 5 or 6 drops of ninhydrin and heat. A blue colouration results.

Ninhydrin gives a colour reaction with amino acids and peptides including proteins. In the solution under test the amino acids with free amino groups are deaminated with the formation of ammonia which reacts with reduced ninhydrin to give a blue or violet colour.

### 3.4.2. BIURET REACTION

*Reagents*—5% sodium hydroxide. Take 5 g sodium hydroxide pellets and dissolve in 20 cm³ distilled water. Cool and make up to 100 cm³.

1% copper sulphate. 1 g copper sulphate made up to 100 cm³.

*Procedure*

To 3 cm³ of solution under test add 3 cm³ of 5% sodium hydroxide and 1 drop of 1% copper sulphate. Shake the solution well. A pink or violet colour should result.

This colour reaction requires the presence of at least two adjacent peptide bonds. The higher polypeptides give deeper colours than the more simple peptides.

### 3.4.2. XANTHOPROTEIC REACTION

*Reagents*—70% nitric acid

Ammonium hydroxide (S.G. 0·880)

*Procedure*

To 3 cm³ of solution under test add 1 cm³ of nitric acid. A white precipitate should form which will turn yellow and partly dissolve to give a yellow solution on heating. Cool and add an excess of ammonium hydroxide. The yellow colouration should deepen to orange if aromatic amino acids are present in the protein.

(An aromatic compound is one which possesses the resonance structure of the benzene ring. The amino acid **tyrosine** is described as an aromatic amino acid.)

Tyrosine

### 3.4.3. SAKAGUCHI REACTION

*Reagents*—5% sodium hydroxide
Industrial alcohol
1% alpha naphthol. 1 g alpha naphthol dissolved in a little industrial alcohol and made up to 100 cm³ with alcohol.
10% sodium hypochlorite

*Procedure*

To 3 cm³ of solution under test add 1 cm³ of 5% sodium hydroxide and 2 drops of 1% alcoholic alpha naphthol. Mix the solution well and then add 1 drop of 10% sodium hypochlorite. A red colour results due to the presence of the amino acid **arginine.** As arginine appears to be a constituent of all proteins this is a good protein test.

### 3.4.4. COLE'S MODIFIED VERSION OF MILLON'S TEST

*Reagents*—Solution A: 100 g mercuric sulphate
100 cm³ concentrated sulphuric acid
Distilled water
*Preparation*—Place 800 cm³ distilled water into a 1 litre flask and slowly pour 100 cm³ sulphuric acid into the water. Keep the flask

cool by holding it under the tap.

Use this diluted acid to dissolve the 100 g of mercuric sulphate and make up the volume to 1 litre (1 dm³).

Filter the solution using a filter pump.

Solution B: 1% aqueous sodium nitrite solution

*Procedure*

To 1 cm³ of the solution under test add 1 cm³ of solution A. Heat gently to boiling and then allow to cool. Note the pale yellow precipitate. Add 0·5 cm³ of solution B and warm gently. A red colouration indicates the presence of the amino acid tyrosine.

## Carbohydrates

Carbohydrates are organic compounds containing carbon, hydrogen and oxygen. Most carbohydrate molecules contain twice as much hydrogen as oxygen.

Soluble carbohydrates are found in the protoplasm of all living cells and their importance lies in the fact that their metabolism is concerned with the distribution and storage of energy.

Some larger carbohydrate molecules have an important structural function, for example, cellulose in plant cells, while others act as carbohydrate reserves, for example, starch in plants and glycogen in animals.

Carbohydrates are traditionally classified according to the length of their molecules.

## Monosaccharides

These are relatively small molecules with 10 or less carbon atoms. This group is subdivided further according to the number of carbon atoms in the molecule.

**Trioses**—3 carbon atoms
**Tetroses**—4 carbon atoms
**Pentoses**—5 carbon atoms
**Hexoses**—6 carbon atoms

*Examples of monosaccharides*

**Ribose** ($C_5H_{10}O_5$). This is a constituent of ribose nucleic acid (R.N.A.) and a constituent of adenosine triphosphate (A.T.P.).

**Glucose**
**Fructose** } All hexoses ($C_6H_{12}O_6$)
**Galactose**

*Properties of monosaccharides*

All monosaccharides are reducing sugars. A reducing sugar is one which will reduce complexes of heavy metals, e.g. copper complexes, to give a precipitate of the free metal or a lower oxide, as in the test using Benedict's reagent.

Monosaccharides are soluble in water.

They are crystalline substances with a sweet taste.

Sugars which have an unsubstituted glycosidic hydroxyl group have reducing properties and form osazones.

$\alpha$ -D - glucose

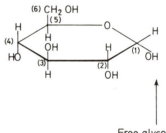

Free glycosidic hydroxyl group

## Disaccharides

These molecules have the general formula $C_{12}H_{22}O_{11}$. A disaccharide is formed from two hexose molecules with the elimination of a molecule of water, an example of a condensation reaction.

$$C_6H_{12}O_6 + C_6H_{12}O_6 \longrightarrow C_{12}H_{22}O_{11} + H_2O$$

Disaccharides may be hydrolysed to release the constituent monosaccharides.

*Examples of disaccharides*

**Lactose**
**Maltose** } Both these molecules contain an active reducing group on one of their constituent monosaccharides and are thus classified as reducing sugars.

α -D-maltose

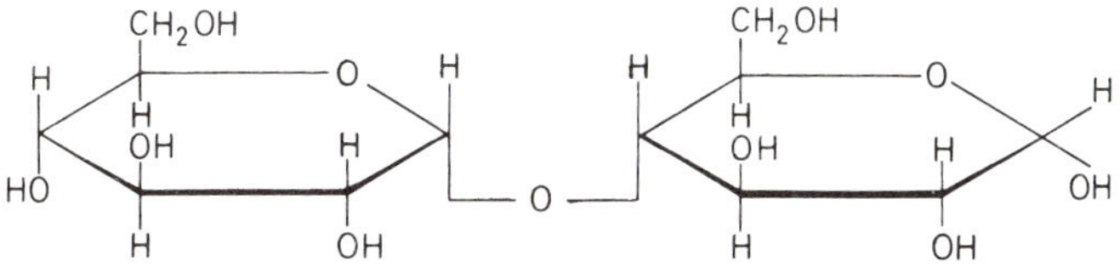

**Sucrose**—In sucrose the two constituent monosaccharides are linked through their reducing groups.

Sucrose

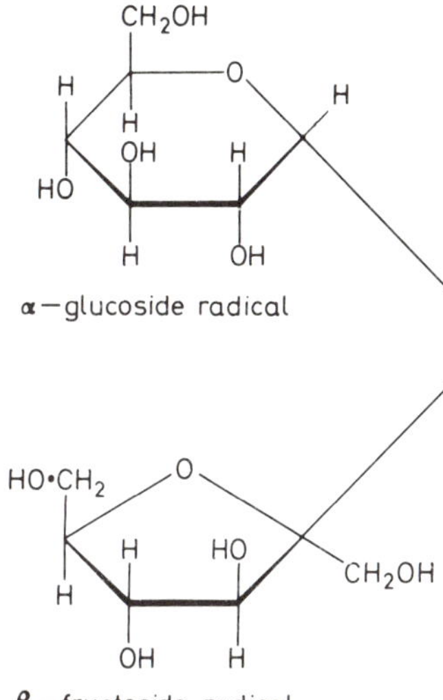

α—glucoside radical

β—fructoside radical

The glycosidic hydroxyl groups of both monosaccharide units are involved in the union to form sucrose, this disaccharide is a non-reducing sugar and will not form an osazone.

## Polysaccharides

Polysaccharides are carbohydrates of high molecular weight. It is not possible to give a definite molecular weight for individual polysaccharides as the size of the molecule and incidentally its detailed structure is characteristic for each species of plant or animal. These larger molecules are formed by the condensation of monosaccharide units.

They have no reducing properties.
They are insoluble in water or form colloidal solutions.
They have no sweet taste.
There are 4 polysaccharides whose names you are likely to encounter.

(a) **Starch**—This molecule is formed entirely from the condensation of **glucose** units. The molecule may be separated into 2 components.

(i) Amylose—This is a spirally coiled molecule consisting of about 300 glucose units. It is soluble in water.

(ii) Amylopectin—This is a branched molecule, each branch having between 20 and 24 glucose units in it. The total number of branches is unknown. It is insoluble in water.

These two components of starch may be separated by allowing the starch to soak in water for several hours.

(b) **Glycogen**—This polysaccharide is only found in animals. It is a condensation product of hundreds of glucose units. The molecule is a highly branched one, each branch containing about 18 glucose units. Its importance as a reserve substance lies in the fact that it may be rapidly hydrolysed to yield glucose when the amount of glucose in the blood falls below the normal level.

(c) **Cellulose**—A condensation product of glucose units linked in long chains. The way in which the glucose units are linked together differs from the arrangement found in the amylose portion of the starch molecule. Mammals have no enzyme capable of splitting the type of linkage found between the glucose units in a cellulose molecule. The digestion of cellulose in herbivores depends on the bacteria found in the alimentary canal of these animals.

(d) **Inulin**—This is a condensation product of fructose units. It is not found in animals but is present as a storage polysaccharide in some members of the Compositae.

## CHEMICAL TESTS FOR CARBOHYDRATES

*Materials required for all tests*
                    Test tubes and test tube holders
                    Test tube racks
                    Bunsen burner
                    Tripod with gauze
                    Beaker or water bath
                    Spatulas

*Suggested carbohydrates for test*

Glucose ⎫
Fructose ⎪
Sucrose ⎬ 0·2% solutions
Maltose ⎭

Starch (Analar soluble starch free of reducing substances)

*Preparation of Starch Solution*—Mix 0·2 g soluble starch with a little cold water taken from 100 cm³ water. Boil the remaining water and then pour in the starch mixture to form a solution.

## 3.5. MOLISCH'S TEST

A positive result is obtained with all carbohydrates.

*Reagents required*—5% alcoholic alpha naphthol
Concentrated sulphuric acid

*Procedure*

To 2 cm³ of the carbohydrate solution under test add a few drops of 5% alcoholic alpha naphthol. Mix. *Very carefully* pour 2 cm³ of concentrated sulphuric acid down the side of the test tube which should be inclined at an angle.

A purple ring develops at the interface of the liquids.

## 3.6. BARFOED'S TEST FOR MONOSACCHARIDES

*Preparation of reagent*—Dissolve 13·3 g copper acetate in 200 cm³ distilled water. Add 1·8 cm³ of glacial acetic acid.

*Procedure*

To 5 cm³ of Barfoed's reagent in a test tube add 0·5 cm³ of glucose solution. Boil for about 30 s. A red precipitate of cuprous oxide should form.

Repeat this test using 0·2% maltose solution and 0·2% sucrose solution in place of the glucose solution. If the solutions are boiled for the same length of time there should be no colour change.


## TESTS FOR REDUCING SUGARS

3.7. BENEDICT'S TEST

*Preparation of Benedict's qualitative reagent*—Dissolve by heating 173 g sodium citrate + 100 g anhydrous sodium carbonate in 800 cm³ water. Cool. Dissolve 17·3 g copper sulphate in 100 cm³ cold water. Add slowly to the first solution and stir well. Make up to 1 litre with distilled water.


*Note*—Benedict's reagent is more stable than Fehling's solution and will keep for long periods in the laboratory. Benedict's reagent is less readily reduced than Fehling's solution and it may be used for detecting monosaccharides in urine. Urine contains uric acid and other substances which will reduce Fehling's solution.


*Procedure*

To 5 cm³ of Benedict's qualitative reagent add 0·5 cm³ of a solution of a reducing sugar and mix well. Boil for 2 min. A red or yellow precipitate of cuprous oxide is produced.


3.8. COLE'S TEST

*Reagents required*—Glycerol
Anhydrous sodium carbonate
5% copper sulphate solution


*Procedure*

To 5 cm³ of the reducing sugar under test add 0·5 g anhydrous sodium carbonate, 3 drops of glycerol and 3 drops of 5% copper sulphate solution. Boil. A yellow-orange precipitate of cuprous oxide is formed.

This test is more sensitive than Benedict's test and with this method as little as 0·01 mg glucose can be detected.

### 3.9. FEHLING'S TEST

*Reagents required*—Solution A:
        Copper sulphate 34·6 g
        Distilled water 500 cm³
      Solution B:
        Sodium hydroxide 77 g
        Sodium potassium tartrate 175 g
        Distilled water 500 cm³

*Procedure*

Add 0·5 cm³ Fehling's solution A to 0·5 cm³ Fehling's solution B. Add this solution to 3 cm³ of the solution under test and boil for a few minutes. Cuprous oxide is precipitated by reducing sugars.

### 3.10. THE HYDROLYSIS OF SUCROSE

*Reagents required*—Benedict's qualitative reagent
        Thymol blue
        Dilute hydrochloric acid
        Solid sodium bicarbonate

*Procedure*

Perform Benedict's test on a 0·2% solution of sucrose. To 3 cm³ of 0·2% sucrose solution add 1 drop of thymol blue and very dilute hydrochloric acid drop by drop until a distinct pink colour is obtained (pH 2·0). Boil the solution for 30 s. Cool. Add a little solid sodium bicarbonate to neutralise the solution. Perform Benedict's test. Note the result and account for it.

### 3.11. SELIWANOFF'S TEST FOR SUGARS CONTAINING A KETONE GROUP

*Seliwanoff's reagent*
        Resorcinol 0·5 g
        Concentrated hydrochloric acid 30 cm³
        Distilled water 30 cm³

*Procedure*

To 1 cm³ of 0·2% fructose solution add a little Seliwanoff's reagent. Heat to boiling. A red colour should quickly develop. (On prolonged heating disaccharides and other monosaccharides will eventually give a red colour.)

## Osazone formation

An osazone is a yellow, crystalline compound formed by heating reducing sugars in a buffered solution with an excess of phenylhydrazine. Glucose, lactose and maltose all form osazones whose crystals group together in characteristic ways.

3.12. THE FORMATION OF GLUCOSAZONE

*Reagents required*—Glucose solution 0·5%
Glacial acetic acid
Phenylhydrazine hydrochloride
Sodium acetate

*Procedure*

To 10 cm³ of 0·5% glucose solution add 1 cm³ glacial acetic acid, 1 g phenylhydrazine hydrochloride and 2 g sodium acetate. (You

Fig. A3. *Diagram to show the arrangement of glucosazone crystals*

may need to warm the tube to dissolve the solids. If it is necessary filter the solution). Put the tube in a boiling water bath for about 20 min and then allow to cool slowly.

A yellow crystalline precipitate of glucosazone should form. Mount some of the crystals in solution on a microscope slide. Examine and draw (*see Fig. A3*).

## TESTS WITH STARCH

### 3.13. IODINE/POTASSIUM IODIDE TEST

*Reagent*—A stock solution containing 10 g iodine and 10 g potassium iodide in 1 litre of water should be diluted to a much paler colour before use.

*Procedure*

To a few cm³ of starch solution add a few drops of dilute iodine/potassium iodide solution. A blue-black colour results. This blue colour which disappears on heating and returns on cooling is due to the formation of a physical adsorption complex between iodine and the amylose component of starch. Amylopectin gives a reddish brown colour with iodine.

You should repeat this test using glycogen instead of starch and note the difference in the colour reaction.

### 3.14. PRECIPITATION OF STARCH WITH ALCOHOL

*Reagents required*—Industrial alcohol
Iodine/potassium iodide solution
Distilled water

*Procedure*

To 2 cm³ starch solution add 8 cm³ alcohol. Mix and allow to stand for 2 min. Filter. Dilute 1 cm³ of filtrate with 2 cm³ water and test with iodine/potassium iodide.

*Further investigation*

From a mixture of carbohydrates, is starch the only one to be precipitated with alcohol?

3.15. ACID HYDROLYSIS OF STARCH

*Reagents required*—N HCl
　　　　　　　　　　Iodine/potassium iodide solution
　　　　　　　　　　Benedict's reagent

*Procedure*

Add 5 cm³ N hydrochloric acid to 20 cm³ of 0·2% starch solution. Mix. Place 5 cm³ of this solution into each of 5 test tubes. Place the tubes in a boiling water bath and note the time. After 1 min remove 1 tube from the bath and cool under a tap. Divide the contents of the tube into 2 equal portions. To one part add iodine/potassium iodide solution dropwise until there is no change in colour. Note the final shade. To the other part add an equal volume of Benedict's reagent and boil. Note the result. (Do not wash out these tubes until the end of the experiment, you will find it helpful to compare the colours of the contents with later results). Repeat the above procedure removing the other tubes after 3, 9, 12 and 15 min. Record your results in tabular form.

## Lipids

Lipids are compounds containing carbon, hydrogen and a relatively small amount of oxygen. A lipid is an ester formed by the reaction between a fatty acid and an aliphatic alcohol. The more complex lipids are derivatives of esters and may contain nitrogen and phosphorus.

Lipids are insoluble in water but are soluble in organic solvents e.g. ether, petroleum and chloroform.

## Simple lipids

This group includes the **fats, oils** and **waxes.** These are all water repellent substances.

**Fats**—A fat is formed from the complete esterification of the alcohol **glycerol.** One molecule of glycerol combines with three molecules of fatty acid. There may be three different fatty acids involved or the three molecules may be of the same fatty acid. For example, when glycerol is esterified with palmitic acid the lipid tripalmitin is formed.

$$
\begin{array}{ccccc}
\text{Glycerol} & \text{Palmitic acid} & \text{Tripalmitin (lipid)} & & \\
CH_2OH & C_{15}H_{31}COOH & CH_2COOC_{15}H_{31} & & \\
| & & | & & \\
CHOH & + \quad C_{15}H_{31}COOH & \longrightarrow \quad CHCOOC_{15}H_{31} & + & 3H_2O \\
| & & | & & \\
CH_2OH & C_{15}H_{31}COOH & CH_2COOC_{15}H_{31} & & \\
\text{Alcohol} & + \quad \text{Acid} & = \quad \text{Ester} & + & \text{water}
\end{array}
$$

The majority of the fatty acids involved in fat formation are **saturated** fatty acids. A saturated fatty acid is one in which all the valency bonds of the carbon atoms in the straight chain are attached to hydrogen atoms.
e.g.

Fats are solid at 20°C.

**Oils**—These are also esters of glycerol but they are liquid at 20°C.
Many oils contain a high percentage of **unsaturated** fatty acids.
An unsaturated fatty acid is one where all the available valency bonds of the carbon atoms are not fully saturated with hydrogen atoms and there is at least one double bond between carbon atoms in the chain.
Part of a chain of an unsaturated fatty acid. e.g.

**Waxes**—Waxes are esters formed from the reaction between higher fatty acids and alcohols of more complex structure than glycerol.

## Compound lipids

Some biologically important compound lipids are those known as **phospholipids** (lipids containing phosphorus). These are essential constituents of the cell membrane. The red blood cell membrane consists almost entirely of a phospholipid called **lecithin**. (Cross reference to Part A2. Further investigation 2.3.) If we take lecithin as one example of a phospholipid you can see from the diagram below that the molecule consists of two main parts with differing fat and water solubility.

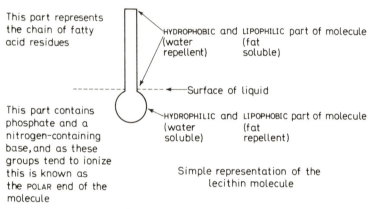

This part represents the chain of fatty acid residues

HYDROPHOBIC and LIPOPHILIC part of molecule
(water            (fat
repellent)        soluble)

Surface of liquid

This part contains phosphate and a nitrogen-containing base, and as these groups tend to ionize this is known as the POLAR end of the molecule

HYDROPHILIC and LIPOPHOBIC part of molecule
(water            (fat
soluble)          repellent)

Simple representation of the lecithin molecule

Phospholipid molecules tend to form films on the surface of liquids with the long axis of the molecules at right angles to the surface of the liquid.

## Sterols

Sterols are lipids with the hydrocarbon chain condensed into a characteristic ring structure. Sterols are monohydroxy alcohols which may be present in the free state or combined with fatty acids as esters.

**Cholesterol,** $C_{27}H_{45}OH$, is an important sterol in lipid metabolism and is present in many animal tissues. It may be synthesised from acetyl groups in the animal body.

It appears that cholesterol is the precursor of the steroid hormones produced in the adrenal cortex. The sex hormones secreted by the gonads have a similar ring structure to that of cholesterol and are classified as steroid hormones. In animal tissues cholesterol may be converted to 7-dehydrocholesterol and when this is present in human skin and exposed to ultra violet light Vitamin $D_3$ is formed.

## CHEMICAL TESTS FOR SOME LIPIDS AND THEIR CONSTITUENTS

*Materials required for all tests*
> Test tubes in test tube racks
> Test tube holders
> Bunsen burner
> Filter paper
> Olive oil
> Glycerol
> Oleic acid

### SOLUBILITY TESTS

*Reagents required*—Ethanol or dry industrial alcohol
Ether

3.16. Shake a drop of olive oil with 5 cm³ ethanol. Warm *cautiously* if the oil does not dissolve. Pour some of the clear solution into water. Some of the oil comes out of the solution as very fine droplets giving an emulsion.

3.17. Add a few drops of olive oil to about 1 cm³ of ether in a test tube. Shake well and note what happens. Pour a drop of this liquid on to a piece of filter paper and leave to dry. Note the translucent patch left after the ether has evaporated.

3.18. REACTION WITH SUDAN IV

*Reagent*—Sudan IV 5 g in 95 cm³ 70% alcohol

*Procedure*

When a few drops of olive oil are shaken up in about 2 cm³ water to form a temporary emulsion and a few drops of Sudan IV are added it will be seen that if the liquid is left to stand the oil which separates out on the top will be stained red by the Sudan IV.


3.19. SAPONIFICATION

*Reagents required*—Oleic acid
                      10% sodium hydroxide
                      Distilled water
                      Concentrated sulphuric acid
                      Sodium chloride
                      Calcium chloride solution

    When fats are hydrolysed they yield glycerol and fatty acids. On hydrolysis with an alkali glycerol is released and the alkaline salts of the fatty acids (soaps) are formed. If the hydrolysis is carried out with sodium or potassium hydroxide the soaps formed are soluble.

    A soap may be formed from the fatty acid, oleic acid. This fatty acid is present in most fats and oils.


*Procedure*

To about 10 drops of oleic acid add 10 cm³ boiling distilled water and then add an excess of 10% sodium hydroxide drop by drop to the boiling mixture until the solution clears. You should now have a solution of soap.

    Divide the solution into 3 parts.

(1) Add a few drops of concentrated sulphuric acid. Leave to stand. The oleic acid should separate out as an upper layer.

(2) Add a little solid sodium chloride and shake. The soap should be precipitated out and rise to the surface.

(3) Add a few drops of calcium chloride solution. Calcium oleate (curd soap) is precipitated out.


## TESTS FOR GLYCEROL

*Reagents required*—Potassium hydrogen sulphate
                      Borax (0·5% solution)
                      Phenolphthalein

Sodium hydroxide (1 % solution)
Copper sulphate solution

3.20. Heat a drop of glycerol with a little powdered potassium hydrogen sulphate. Note the pungent smell of **acrolein.**

3.21. DUNSTAN'S TEST
To 5 cm³ 0·5% borax solution add enough phenolphthalein to give a marked pink colour. Add glycerol *drop by drop* until the pink colour disappears. Boil. The pink colour will return.

3.22. To 2 drops of glycerol add about 1 cm³ of 1% sodium hydroxide solution and a few drops of copper sulphate solution. A deep blue solution is obtained, glycerol preventing the precipitation of cupric hydroxide.

## Vitamins

Vitamins are organic compounds. They do not belong to any one class of organic compound and differ widely in their chemical structure. For example, vitamin A is an alcohol, vitamin $B_1$ is the hydrochloride of an organic base while vitamin C is an acid, ascorbic acid.

Vitamins are necessary in small amounts for the healthy functioning of an animal. They are essential for metabolism and many have been shown to be parts of enzyme systems only active in a limited range of temperature and pH.

It is not possible to make any generalisations about vitamin manufacture as different species of animals vary in their ability to synthesise vitamins.

Most school advanced biology textbooks list the names, sources and functions of those vitamins which have been isolated and investigated, such descriptions are beyond the scope of this book.

With school laboratory facilities it is possible to demonstrate the presence of vitamin A and vitamin C.

3.23. TO DEMONSTRATE THE PRESENCE OF VITAMIN A

*Reagents required*—Vitamin A source (cod liver oil or halibut oil)
                              Chloroform
                              Saturated solution of antimony trichloride in
                                   chloroform

*Procedure*

To 1 cm³ of cod liver oil add 5 cm³ chloroform. To a drop of this solution add a few drops of a saturated solution of antimony trichloride in chloroform. An intense blue colour forms.

3.24. TO DEMONSTRATE THE PRESENCE OF VITAMIN C

*Reagents required*—0·1% aqueous ascorbic acid solution
0·1% aqueous DCPIP solution (dichlorophe-nol-indophenol) prepared by dissolving 1 g in 1 litre of water. This should be freshly prepared.

*Procedure*

To 1 cm³ of DCPIP solution add ascorbic acid solution drop by drop. The blue dye turns pink and the colour fades. This is due to the powerful reducing action of ascorbic acid.

## 4. THE BEHAVIOUR OF COLLOIDS IN SOLUTION

Thomas Graham working between 1851–1861 was the first person to use the term **colloid** to describe those substances which were not crystalline in the solid state and which only diffused slowly in solution. Crystalline substances which diffused rapidly in solution he named the **crystalloids.**

We know now that it is not possible to divide rigidly all substances into these two groups as some of the substances formerly classified as crystalloids may be prepared in the colloidal state.

A substance is described as colloidal if the diameter of a particle lies in the range 1 μm − 1 nm (1 μm = $10^{-6}$ metre, 1 nm = $10^{-9}$ metre). A particle may be a single large molecule, e.g. a protein molecule, or a particle may be formed from a group of smaller molecules as in the case of some inorganic substances.

A solution containing particles in this size range is known as a colloidal **sol.** The particles form the **disperse phase** and the liquid in which they are dispersed is known as the **continuous phase** or **dispersion medium.** A sol may be converted to a semi solid **gel** and when this occurs the particles form a network, that is, they form the continuous phase with the liquids trapped in the spaces and described now as the disperse phase.

When blood clots there is a change from the sol to the gel state. Plasma is a colloidal sol with fibrinogen, one of the plasma proteins, dispersed throughout the liquid part of the plasma. When a clot or gel forms fibrinogen is converted to a meshwork of fibrin forming the continuous phase in which the plasma is dispersed. This change is irreversible.

In some cases the sol to gel change is reversible.

The colloidal substances e.g. protein and polysaccharide molecules, found in living cells show an affinity for water and are described as **hydrophilic colloids.** In solution there is an interaction between water molecules and the colloid particles which tends to stabilise the colloid and prevent it from being precipitated by small amounts of electrolyte.

A colloidal solution has a very small **osmotic potential** when compared with a true solution (solute of diameter below 1nm) containing the same weight of solute. The colloidal solution will have a relatively small number of molecules compared with the true solution and the osmotic potential is proportional to the number of solute molecules per unit volume (*see Part A2*).

Graham showed that when a solution of a colloid and a solution of a crystalloid are placed together in a parchment bag suspended in water the crystalloid will diffuse through the parchment membrane while the colloid is left behind in the bag. This method of separation is known as **dialysis.** Separation of the two substances is possible because of the different molecular size of the crystalloid and colloid. Any membrane with a pore size smaller than the diameter of the colloid but which will allow the passage of the crystalloid will function as a dialysing membrane.

It is possible to use Visking tubing as the dialysing membrane in the separation of the crystalloids and colloids found in milk.

4.1. DEMONSTRATION OF DIALYSIS

*Materials required for each demonstration*
>35 cm length of 2·5 cm diameter Visking tubing
(If Visking tubing of smaller diameter is used the volume of milk should be adjusted accordingly. Visking tubing of different diameters may be obtained from Griffin & George, Dutt, Philip Harris & Gerrards)
>900 cm³ capacity gas jar
>Retort stand and clamps

(Motorised stirrer—optional) or
Glass rod to act as stirrer
35 cm length of thread
Bunsen burner
Test tubes in rack
Two 5 cm³ graduated pipettes
50 cm³ milk
500 cm³ water
Benedict's reagent (*see Part A3*)
Solution A and Solution B of Cole's modified
    version of Millon's test (*see Part A3*)
10% trichloroacetic acid (*see Part A3*)

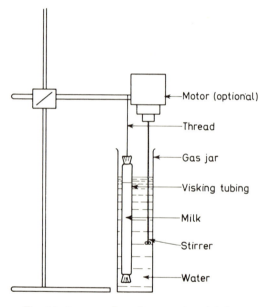

Fig. A4. *Apparatus for a demonstration of dialysis*

*Procedure*

Soak a 14 in length of Visking tubing in water for a few minutes. Tie
a double knot in one end and then pour 30 cm³ of milk into the open
end. Leave a space of about 2 in above the milk before tying a knot
in the top end of the tubing. Tie one end of a length of thread underneath the top knot and suspend the tubing in a gas jar containing
water. If you have a motorised stirrer place it in the gas jar by the side
of the Visking tubing. Start the motor and leave for about 2 h.

While dialysis is taking place you should perform the following tests on samples of the milk from which the 30 cm³ were taken.
(1) Benedict's test for reducing sugars. Perform a control test using water in place of the milk sample and note any difference in results.
(2) Cole's modification of Millon's test for the detection of proteins. Perform a control test using water in place of the milk sample and note any difference in results.

The results of these two tests on milk samples should show that milk contains both reducing sugars and proteins.

Leave your test tubes in the rack for comparison with later results.

*After 2 h or longer*—Remove samples of the liquid in the gas jar outside the Visking tubing and perform the following tests.
(1) Benedict's test.

If you get a negative result with this test try Cole's test for reducing sugars (*see Part A3.8*, page 20).
(2) Cole's modification of Millon's test.
(3) To 3 cm³ of the solution add 3 cm³ of 10% aqueous trichloro-acetic acid (Caution—prevent contact of this reagent with the skin). If there is any protein present at all the solution will become turbid and a white precipitate will develop.

*Interpretation*—A positive result for (1) and negative results for (2) and (3) indicate that the Visking tubing used in the experiment is permeable to reducing sugars but is not permeable to the proteins found in milk, it is thus described as **differentially permeable** and behaves as a dialysing membrane.

*Further work*—Name the organs in the mammalian body where dialysis occurs and list the substances which pass through the differentially permeable membranes in these organs.

## Brownian movement

Named after Robert Brown an English Botanist, who recorded having seen this type of movement.

It is a vibratory movement shown by particles of both inorganic and organic substances small enough to be moved as a result of the bombardment by the constant vibrations of surrounding molecules.

The particles in a colloidal sol show Brownian movement.

## 4.2. DEMONSTRATION OF BROWNIAN MOVEMENT

*Materials required*—Milk
      Teat pipette
      Microscope slide and cover slip
      Microscope


*Procedure*

Dilute 1 drop of homogenised milk with 1 drop of water. Place a drop on a microscope slide and cover with a cover slip. Move the slide until a small globule of fat is visible. Watch carefully using a microscope to observe any quivering movements.

# Enzymes

Living organisms produce enzymes which behave as catalysts acting on one or more specific substrates. All the enzymes which have been isolated and purified are proteins or contain a protein as an essential component. Every stage of the many reactions comprising metabolism is catalysed by enzymes. The names of all enzymes indicate the type of reaction that each enzyme will catalyse. The compound upon which an enzyme acts is called the **substrate** and the type of substrate acted upon is often indicated in the name of the enzyme. For example a **lipase** acts on a lipid and a **protease** on a protein. Examples of respiratory enzymes which are named according to the type of reaction catalysed are the **oxidases** which catalyse oxidations in the presence of free oxygen and the **dehydrogenases** which are active in reactions involving the transference of hydrogen from a substrate to a hydrogen acceptor.

As a catalyst an enzyme causes an increase in the velocity of a chemical reaction. A very small amount of enzyme may catalyse relatively large amounts of reactants. An enzyme is largely restored at the end of the reaction. The activity of an enzyme varies with temperature, pH of the medium and with the proportions of the reactants.

Some enzymes are highly specific and will catalyse only one reaction, e.g. **urease** will only catalyse the breakdown of urea to carbon dioxide and ammonia. Other enzymes are less specific and may catalyse a group of substrates rather than a single substrate, e.g. pancreatic lipase catalyses the hydrolyses of a range of fats.

In order to be active many enzymes need the presence of accessory substances. In some cases the activator is an inorganic ion. Salivary **amylase** (ptyalin) requires $Cl^-$ as an activating factor and $Ca^{++}$ is one of the factors necessary for the activity of **thrombase.**

Enzyme activity is inhibited by various organic and inorganic substances especially those which act as protein precipitants.

*Experimental procedure for enzyme experiments*

Many enzymes may be obtained in powder form and it is possible to demonstrate their catalytic activity in the laboratory.

It is convenient to have a water bath maintained at a temperature of 37°–38°C.

With all enzyme experiments a control using a boiled enzyme should be included.

All tubes should be clearly labelled before they are placed in a water bath. Pencils for labelling glassware are obtainable from suppliers of scientific apparatus.

# 1. ACTIVITY OF DIGESTIVE ENZYMES

During digestion complex organic molecules are broken down to more simple molecules which may pass through the lining of the small intestine. The breakdown of complex molecules is achieved by a series of hydrolyses (Greek—*hydor* water, *lyein* to dissolve) catalysed by the digestive enzymes produced by the glandular cells of the digestive system.

1.1 HYDROLYSIS OF STARCH IN THE PRESENCE OF SALIVARY AMYLASE

Salivary amylase (ptyalin) is the enzyme found in saliva, it begins the hydrolysis of cooked starch. It is activated by chloride ions and has an optimum pH of 6·8–6·9.

Amylase activity in a starch and amylase mixture may be demonstrated by the disappearance of starch and the appearance of a reducing sugar.

*Materials required*—Analar soluble starch
                Saliva
                Sodium chloride
                Iodine in potassium iodide in dropping bottle
                  (*see Part A3.13*)
                Benedict's solution (*see Part A3.7*)
                Barfoed's reagent (*see Part A3.6*)
                Distilled water

Test tubes in rack
1 measuring cylinder (100 cm³)
Bunsen burner, tripod, gauze
Water bath, maintained at 37°C
One 10 cm³ graduated pipette
Glass rods
Two 250 cm³ beakers
Spatula
2 pipettes with teats

*Preparation of starch solution*—Mix 0·5 g starch to a thin paste with a little cold water taken from 100 cm³. Boil the rest of the water and add 50 mg sodium chloride. Pour this solution on to the starch paste, stirring all the time. Leave to cool to 37°C. Test a few cm³ with Benedict's solution to make certain that the starch is free from reducing sugars.

*Preparation of ptyalin solution*—Chew a piece of rubber tubing for a few minutes until you start to produce a good flow of saliva. Collect some saliva in a beaker and dilute it with an equal volume of distilled water. Place the beaker in a water bath so that the saliva will attain a temperature of 37°C.

*Procedure*

Pour a small amount of saliva solution into each of two test tubes. Heat one tube to boil the saliva solution, this will act as the control tube.

Add 10 cm³ of starch solution to both tubes and add 1 drop of iodine/potassium iodide to each tube. Note the colouration in each. Place both tubes in the water bath at 37°C. Examine both tubes every 5 min for at least 30 min and note any changes in colouration.

Use a separate pipette for each solution to withdraw a few drops from each tube and perform Benedict's test.

If you have a positive reaction from Benedict's test, perform Barfoed's test on a sample of the solution, being careful to boil the solution for 30 s only.

What can you deduce from your results about the hydrolysis of starch?

1.2. HYDROLYSIS OF EGG ALBUMEN BY A PROTEASE (PEPSIN)

Egg albumen (Latin—*albumen* white of egg) contains an albumin, a type of protein found in a variety of biological tissues. Pepsin is a protease which begins the hydrolysis of proteins in the stomach.

*Materials required*—1 egg white
              Pepsin powder
              Hydrochloric acid, 0·1 M
              Test tubes in rack
              Muslin or glass wool
              Glass rod
              Two 10 cm³ graduated pipettes
              Two 250 cm³ beakers
              1 filter funnel to fit a 250 cm³ beaker
              One 1 litre beaker (to act as a water bath at
                60°C)
              Bunsen burner, tripod and gauze
              Water bath—maintained at 37°C

*Preparation of egg albumen suspension*—Thoroughly stir the egg white in a beaker and then dilute it with 4 times its own volume of water. Place the beaker in the 1 litre beaker of water maintained at 60°C. Leave for about 20 min. Strain the albumen suspension through muslin or glass wool. A suitable concentration of the suspension to use is made by diluting 1 part of albumen with 9 parts water.

*Preparation of pepsin solution* (1 %)—Dissolve 1 g pepsin powder in 100 cm³ water.

*Procedure*

Pipette 1 cm³ of 1 % pepsin solution into each of two test tubes. Heat one tube to boil the pepsin solution, this will act as the control tube.

In each tube place 5 cm³ egg albumen suspension together with 5 cm³ 0·1 M hydrochloric acid. The acid is necessary to achieve a suitable pH for pepsin activity.

Place both tubes in a water bath at 37°C and leave for 45 min.

Compare the appearance of the contents of both tubes after this time.

1.3. HYDROLYSIS OF CASTOR OIL BY LIPASE

A lipase is an enzyme which catalyses the hydrolysis of fats and oils to diglycerides, monoglycerides and fatty acids.

*Materials required*—Castor oil
Lipase powder
Bromothymol blue
Alcohol, 50%
Two 10 cm³ graduated pipettes
Water bath, maintained at 37°C
Test tubes in rack
Bunsen burner

Bromothymol blue is the indicator used here to detect the change in pH during the course of the hydrolysis. The range of this indicator is from pH 6·0 to pH 7·6 and the colour variation is shown below.

pH 6·0—yellow
(neutral) pH 7·0—green
pH 7·6—blue

*Procedure*

Into each of 3 test tubes pipette 1 cm³ castor oil and 2 cm³ 50% alcohol. Shake the tubes well.

To tube (1) add 0·1 g lipase powder (or 1 cm³ of 1% lipase solution). Shake the tube well.

To tube (2) add 0·1 g lipase powder (or 1 cm³ of 1% lipase solution). Shake the tube well and then heat to boiling. This is the control tube.

To tube (3) add a few drops of a neutral solution of bromothymol blue and note the colour. Leave the tube as a colour standard for later reference.

Place tubes (1) and (2) in the water bath at 37°C and leave for 30 min.

After 30 min remove a few drops from each tube in turn and add to a little bromothymol blue (neutral solution). Note any colour change and continue to extract samples at 15 min intervals until a distinct yellow colour is produced by the contents from tube (1).

*If time allows*—Test for the presence of glycerol in the contents of tubes (1) and (2) (*see Part A3.20–3.22*).

1.4. FURTHER INVESTIGATION

*Materials required*—A small slice of fresh liver (about 1 oz)
                    Sodium chloride ($0.9\%$ solution)
                    Distilled water
                    Milk
                    Litmus solution
                    1 measuring cylinder (100 cm³)
                    Clean sand
                    Scalpel or knife
                    Pestle and mortar
                    Two 250 cm³ beakers
                    Filter funnel and filter paper or centrifuge
                    Glass rod
                    Hypodermic syringe (2 cm³)
                    Test tubes in rack
                    Water bath maintained at 37°C

*Preparation of liver extract*—Chop 1 small slice of fresh liver into pieces and wash in $0.9\%$ sodium chloride solution to extract blood.

Remove the small pieces of liver and grind with clean sand in a mortar. Tip this mixture into a beaker and add 100 cm³ distilled water. Stir well and allow to stand. Decant the liquid and filter or centrifuge.

Use the filtrate or supernatant.

*Procedure*

One-third fill 3 test tubes with filtrate/supernatant.
Boil the contents of 1 test tube and label the tube, CONTROL.
Add enough milk to each of the 3 test tubes until they are two-thirds full.
With the test tubes in a rack in front of a white background add by syringe approximately 20–25 drops of litmus solution to each tube.
Shake the tubes well.
Contents of the tubes should appear *just* blue.
Leave 1 test tube in the rack.
Place the other tube together with the control tube in the water bath maintained at 37°C.
Examine the tubes after 1 h and carefully compare the colours of all three tubes.

Can you formulate a hypothesis to account for the results of this experiment?

## 2. ACTIVITY OF OXIDISING AND REDUCING ENZYMES

These are the enzymes catalysing the oxidations and reductions of tissue respiration.

Tissue or internal respiration is the name given to the sum of reactions which result in the release of energy from the breakdown of chemical compounds within the cells of an organism. Biological oxidation of a substrate generally takes place by dehydrogenation, the removal of hydrogen from the substrate molecule. Hydrogen is transferred from the substrate to a hydrogen acceptor and after passing along a chain of acceptors is finally removed as water by oxidation with oxygen which is supplied to the cells. It is during the process of dehydrogenation that energy is released and used to form the high energy compound adenosine tri-phosphate, A.T.P. The importance of this molecule is that it holds energy in a form which is readily available to the cell. When energy is used by an organism, as for example in muscular contraction or in biosynthetic processes the energy is supplied by the breakdown of A.T.P.

2.1. TO DETERMINE THE PRESENCE OF A DEHYDROGENASE IN MUSCLE

A dehydrogenase is an enzyme catalysing a reaction in which hydrogen is transferred from one substrate to a hydrogen acceptor. As a result of the reaction the substrate is oxidised and the hydrogen acceptor is reduced.

i.e.

$$\underset{\substack{\text{Substrate}}}{AH_2} + \underset{\substack{\text{Hydrogen}\\\text{acceptor}}}{B} \xrightarrow[\text{dehydrogenase}]{\substack{\text{in the presence}\\\text{of a}}} \underset{\substack{\text{Oxidised}\\\text{substrate}}}{A} + \underset{\substack{\text{Reduced}\\\text{hydrogen}\\\text{acceptor}}}{BH_2}$$

In this demonstration a dilute solution of the dye, methylene blue is used to show the presence of a dehydrogenase. In the reduced form methylene blue loses its colour.

*Materials required*—Small pieces of muscle from a freshly killed animal
    Methylene blue (0·03% solution)
    Scalpel
    Distilled water
    Two 1 cm³ graduated pipettes
    2 Thunberg tubes

1 test tube
Vacuum pump
Water bath at 37°C

*Procedure*

Remove fat and connective tissue from the muscle and cut the muscle into small pieces with the scalpel.

Place a few pieces of the muscle in a test tube and cover with distilled water. Heat the tube to boil the contents.

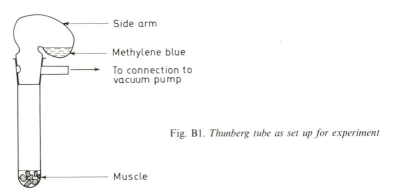

Side arm

Methylene blue

To connection to vacuum pump

Fig. B1. *Thunberg tube as set up for experiment*

Muscle

Pour 2 cm³ of the water and boiled muscle into the main part of one Thunberg tube.

Into the second Thunberg tube place the same amount of unboiled muscle and add about 2 cm³ water.

Pipette 0·5 cm³ of 0·03% methylene blue solution into the side arm of each Thunberg tube.

Connect the side arm of each tube in turn to a vacuum pump to evacuate the tubes. Then rotate the side arm to prevent the re-entry of air and disconnect the vacuum pump.

Place the two tubes in a water bath at 37°C.

Tilt the side arm of each tube to pour the methylene blue solution into the main part of the tube.

Note the time taken for the dye to lose its colour.

2.2. DEMONSTRATION OF THE ACTION OF A BLOOD PEROXIDASE (catalase)

Peroxidases are enzymes present in all tissues which use oxygen for respiration i.e. all aerobic tissues. During the process of tissue respiration hydrogen peroxide is formed but this is immediately decomposed to water and oxygen in the presence of a peroxidase.

The reactions involved may be represented simply thus:—

(1) $$\underset{\text{Substrate}}{AH_2} + \underset{\text{Oxygen}}{O_2} \xrightarrow[\text{an oxidase}]{\text{in the presence of}} \underset{\text{Oxidised substrate}}{A} + \underset{\text{Hydrogen peroxide}}{H_2O_2}$$

(2) $$2H_2O_2 \xrightarrow[\text{a peroxidase}]{\text{in the presence of}} 2H_2O + O_2$$

*Materials required*—Test tube and bung to fit
     Splint
     10 vol hydrogen peroxide solution
     Source of blood: (1) Sterile lancets necessary if blood is to be obtained from the finger, 1 drop of blood will contain enough enzyme for 1 demonstration. (2) Fresh liver as a source of blood peroxidase. Chop up a small piece of fresh liver to release the blood. Dilute with distilled water which will cause haemolysis and the liberation of the enzymes from the blood cells. Filter or centrifuge to remove the tissues and use the filtrate or supernatant

*Procedure*

Threequarters fill a test tube with 10 vol hydrogen peroxide. Add a few drops of blood to the tube. Note the effervescence and quickly place a bung in the mouth of the test tube to trap the evolved gas. Have a glowing splint ready to plunge into the tube on removal of the bung.

## 3. FACTORS AFFECTING THE RATE OF ENZYME ACTION

Three factors will be investigated in three experiments:
     the effect of temperature,
     the effect of enzyme concentration,
     the effect of pH.
 Do not limit your experimental work to the 3 experiments given here. Use the experimental procedure given in B1 and B2 as a basis

and devise your own methods for investigating the effect of all or some of the above three factors on the course of these reactions.

You could then be more ambitious and examine the effects of combined factors, e.g.

Does a relatively low temperature produce the same result at different pH values?

### 3.1.  THE INFLUENCE OF TEMPERATURE ON THE RATE OF THE HYDROLYSIS OF STARCH

*Materials required*—Starch solution $\left.\right\}$ *See Part B1.1* for preparation
Saliva solution $\left.\right\}$ of solutions
Iodine in potassium iodide in dropping bottle
Benedict's solution
12 test tubes in rack
Two 10 cm$^3$ graduated pipettes
Four 250 cm$^3$ beakers
4 thermometers
4 pipettes with rubber teats
4 bunsen burners, tripods and gauzes

*Procedure*

Maintain 4 beakers of water at 20°C, 30°C, 40°C and 50°C.

Place 3 cm$^3$ starch solution in each of 8 test tubes. Place 2 tubes in each beaker and leave for a few minutes until the starch solution reaches the temperature of the water bath.

*Controls*—Boil about 4 cm$^3$ saliva solution in a test tube and then add the same amount to one of the tubes in each water bath. Label these tubes.

Add an equal amount of unboiled saliva solution to the second tube in each water bath.

Add 1 drop iodine/potassium iodide to give a distinct colouration in each tube. Note the time.

At 5 min intervals note any difference of colour in each tube.

Perform Benedict's test on a few drops of solution from each tube to determine the presence of any reducing sugars.

Record the time taken for the starch to be hydrolysed at each of the 4 temperatures.

Plot a graph of the time taken for the starch to be hydrolysed

($y$ axis) against the exact temperature of each water bath ($x$ axis). Does the rate of reaction double for every 10°C rise?

### 3.2. THE EFFECT OF ENZYME CONCENTRATION ON THE HYDROLYSIS OF STARCH BY DIASTASE

*Materials required*—Starch solution—prepared as in B1.1 but using
0·2 g starch instead of 0·5 g starch
Diastase solution—2·0% (free from reducing sugars). Prepare diastase solutions from the stock 2% solution in the following concentrations: 1%, 0·5%, 0·25%
Iodine in potassium iodide—diluted to pale straw colour
Benedict's reagent
12 test tubes in rack
2 graduated pipettes (10 cm³)
Water bath at 38°C
4 pipettes with rubber teats
Glass rod

*Control*—Boil 2 cm³ 2% diastase solution.

*Procedure*

Place 2 cm³ of 0·2% starch solution in each of 5 labelled test tubes. Place the tubes in the water bath to attain a temperature of 38°C. Then add 2 cm³ of diastase solution of the following concentrations:
Tube (1) 2%
Tube (2) 1%
Tube (3) 0·5%
Tube (4) 0·25%
Tube (5) (Control) 2% boiled diastase
To each tube add *1 drop* of iodine to give a deep blue colouration. Mix the contents of each tube with a clean glass rod.
Record the time taken for the blue colour to disappear indicating that the starch has been hydrolysed.
Perform Benedict's test for reducing sugars on a sample from a tube whose contents are colourless. (Remember first to test both the starch and diastase solutions to show that they are free of reducing sugars.)

Plot a graph of time taken from hydrolysis against enzyme concentration.

### 3.3. THE EFFECT OF HYDROGEN ION CONCENTRATION ON THE HYDROLYSIS OF GELATIN BY TRYPSIN

The gelatin of exposed film is used as the source of protein. When the gelatin is hydrolysed the black particles of silver are freed from the film so that the end point of the experiment is reached when the film appears transparent.

*Materials required*—Pieces of exposed film
0·5% trypsin solution
Tablets for making buffer solutions (supplied by Griffin and George) or
B.D.H. Universal buffer solution
0·2 N sodium hydroxide
0·2 N hydrochloric acid
Two 250 cm³ measuring cylinders
Four 250 cm³ beakers
Universal indicator papers (pH 1·0–10·0)
8 splints
Water bath at 38°C
9 test tubes
2 graduated pipettes (1 cm³)

*Preparation*—Buffer solutions of pH 4, pH 6, pH 8 and pH 10 should be prepared either using the tablets or the universal buffer solution.

*Instructions for the preparation of solutions of known pH from the universal buffer solution (Taken from the B.D.H. leaflet)*—To 100 cm³ of solution add the amount of acid or alkali necessary to produce the required pH and dilute with water to 200 cm³.
$$pH = 3·1 \pm 0·1185v$$
where $v$ = cm³ of 0·2N NaOH or HCl and NaOH is (+) and HCl is (−), i.e. the addition of each cm³ of 0·2N NaOH or HCl will alter the pH by ± 0·1185 respectively.

Each buffer solution should be tested with a universal indicator paper.

*Control*—Boil 1·0 cm³ 0·5% trypsin solution.

*Procedure*

Take 4 pairs of labelled test tubes and label 1 tube of each pair control.

Add 0·2 cm³ buffer solution to each tube as follows:—

> 1st pair pH 4
> 2nd pair pH 6
> 3rd pair pH 8
> 4th pair pH 10

To the control tube of each pair add 0·2 cm³ of boiled 0·5% trypsin solution and to the other 4 tubes add 0·2 cm³ of unboiled trypsin solution.

Place all 8 tubes in the water bath and leave for 5 min.

Cut 8 small rectangles of exposed film and fix each one into the split end of a wooden splint.

Then simultaneously place the piece of film attached to a splint in each of the 8 tubes and note the time of immersion.

Record the time taken for the pieces of film to become transparent.

Plot a graph of time for hydrolysis against pH.

## PART C
# Microscopical Preparations

## 1. NOTES ON THE PROCESS OF MAKING PERMANENT STAINED PREPARATIONS

A permanent preparation should remain suitable for microscopic examination for many years. The different stages involved in making such a preparation are outlined in sequence below.

### 1.1. FIXATION

The object to be stained is immersed in a fluid known as a **fixative.** A fixative is satisfactory if it fulfils the following requirements:—
- (*a*) It preserves a tissue with the minimum of distortion.
- (*b*) It hardens the tissue and enables it to withstand subsequent processes.
- (*c*) It prepares the tissues for staining.

In some cases the fixative may be used as a killing fluid. In cases where other methods of killing are employed fixation should be carried out directly the animal has been killed.

#### EXAMPLES OF FIXATIVES

**Acetic-alcohol—Clarke's fluid**

*Reagents*—Glacial acetic acid 25 cm³
100% ethanol* 75 cm³ (*Industrial methylated spirits may be substituted provided that all traces of water have been removed by shaking the alcohol with anhydrous copper sulphate and preventing exposure to moist air.)

A rapidly acting fixative. Good for squashes and cytological work. Hardening properties are not good and it should not be used when fixation will be followed by embedding and sectioning.

### 70% alcohol

Rapidly acting fixative with a good hardening effect.
Disadvantage is that it causes great shrinkage of protoplasm.

### 4% formaldehyde

Relatively slow fixative with great hardening properties.
Disadvantage is that it makes staining with acid dyes difficult.

### Formol-saline

4% solution of formaldehyde in isotonic saline.
A good but slow acting fixative for general routine work.

### Bouin's fluid

*Reagents*—Glacial acetic acid 5 cm³
          40% formaldehyde 25 cm³
          Picric acid (saturated aqueous 75 cm³)
    Rapid fixative with good penetration. No great hardening properties. Shrinkage of cytoplasm minimal.

*After fixation tissues should be washed well to avoid interference with subsequent processes.*
    Use 100% alcohol to wash off 'acetic alcohol'. 70% alcohol should be used for the remaining fixatives listed above.
    For preparations of small whole animals, e.g. protozoans, small arthropods, embryos or larval forms proceed to 1.5.

## TECHNIQUES EMPLOYED TO PREPARE TISSUE FOR STAINING

### 1.2. WAX IMPREGNATION, EMBEDDING AND SECTIONING

These processes are necessary if very thin sections of animal tissues are required for staining.

If after fixation the tissue was washed in 70% alcohol it should now be transferred to 90% alcohol and then into two changes of absolute alcohol to dehydrate the tissue.

The dehydrated tissue should then be transferred to a clearing agent, either chloroform, benzene, xylene or cedarwood oil.

### WAX IMPREGNATION

Immerse the object in a mixture of equal parts clearing medium and paraffin wax (M.P. 53°C) in an oven (temperature not higher than 58°C) for 15–30 min.

Transfer the object to pure molten wax for 30–60 min according to size.

With a warm section lifter, transfer the object to a second bath of wax in the oven for 30–60 min. After this time all the clearing agent should be replaced by wax.

### EMBEDDING

The wax impregnated object must now be embedded in a block of paraffin wax.

Smear a container, which may be a solid watch glass or a rectangular mould made from two L-shaped pieces of brass on a glass plate, with a little glycerine and pour in fresh molten wax. The wax should

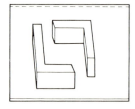

Fig. C1. *Rectangular mould for wax embedding*

be allowed to set slightly at the bottom of the mould to supply support for the object. Add the latter with a warm section lifter and arrange it in the best position for cutting sections. Add more molten wax if necessary. Blow gently on the surface of the wax to form a skin and when the wax is sufficiently solid rapidly cool the block by immersing it in cold water.

After the wax has completely solidified it can be removed from the mould and trimmed for sectioning.

SECTIONING

With a mechanical microtome it is possible to cut a ribbon of sections of known thickness.

The section thickness gauge should be set to a value between 8–12 μm (micrometres). 1 μm (1 micrometre) = 0·001 mm.

The trimmed wax block must be firmly fixed to the chuck of the microtome. The handle of an old scalpel can be heated and used to melt the wax on the chuck. The wax block is then applied to the melted wax and the hot scalpel handle can be used to smooth the junction of the block and chuck. Hold the chuck and attached block under cold water for a few minutes. Then fix the chuck to the microtome ready for section cutting.

Loose sections may become damaged or even lost. It is best to attach them to a slide or cover slip before embarking on a staining routine.

ATTACHMENT OF SECTIONS TO THE SLIDE

The adhesive used for attaching sections to slides may be a mixture of glycerine and egg albumen solution (3 drops of glycerine and albumen to 25 cm³ distilled water) or human saliva.

Pipette sufficient adhesive on to a clean slide to allow the section(s) to float freely.

Warm the slide gently until the wax is fully expanded.

*Do not melt the wax.*

Draw off excess water with a filter paper.

Allow to dry thoroughly.

Remove the wax with xylene.

Transfer to a mixture of 1 part xylene and 1 part 'absolute alcohol' (undiluted industrial meths).

Remove all traces of xylene by immersing in 'absolute alcohol'.

Proceed to 1.5.

1.3. TEASING OR DISSOCIATION OF CELLS

This is the process used to separate out the cells of a tissue from one another to make easier the examination of their individual structure.

The tissue may be teased apart with two glass needles while it is in the fixative. Adequate teasing is essential to ensure penetration by the fixative.

For preparations of striated muscle, nerve or tendon, the fibres may be fixed and teased apart in 70% alcohol.

## 1.4. SMEARING

A semi liquid or liquid tissue is best examined as a smear. A smear should be a *very thin film uniformly distributed* over the slide or coverslip (*see Part C2*).

## 1.5. STAINING

The purpose of staining an object is to enable its different components to be seen more clearly than in the unstained state. If only one stain is used, to be satisfactory, it should be taken up in differing degrees of intensity by the different components.

More than one stain may be used with the object of staining the different components in contrasting colours, this method is known as **counter staining.**

*Important points to remember before staining*

(1) The solvent of the stain should be known and if the stain needs to be diluted before use, dilute it with the solvent.
(2) If the solvent is alcohol remember to dehydrate the fixed and washed tissue up to that strength of alcohol.

### USEFUL STAINS FOR ANIMAL TISSUES

(1) **Borax carmine** (solvent, 50% alcohol)
A good general stain for small, whole invertebrates.

(2) **Mayer's acid haemalum** (solvent, distilled water)
Good stain for sections, small whole animals, bulk material and nuclei.

(3) **Ehrlich's haematoxylin** (solvents equivalent to 70% alcohol)
A stain suitable for general animal histology.
May be counterstained with eosin.

**(4) Leishman's** (solvent, methyl alcohol)
A stain for blood and blood parasites.

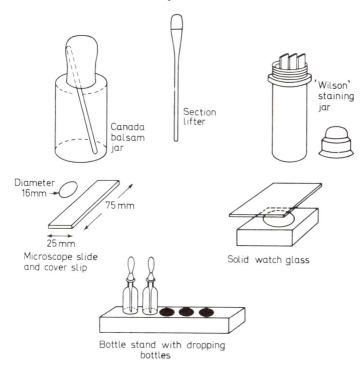

Fig. C2. *Type of equipment needed for making microscopic preparations*

TECHNIQUE OF STAINING

The object to be stained (unless it is in the form of a smear or a section attached to a slide) should be placed in a solid watch glass containing enough stain to completely cover the object. When transferring the object use a section lifter. If you use forceps or a needle there is a danger that you may tear the tissue and spoil the preparation.

You may find it necessary to make some adjustments to the times for staining given in the individual schedules. Examine the tissue being stained under the low power of a microscope after the minimum time suggested in the schedule. If you need to leave the tissue longer in the stain examine it frequently to prevent overstaining. It is easier for a beginner to understain a tissue at first and then return it to the stain so that more may be taken up (progressive staining) rather than to overstain to begin with and have to remove some of the stain later (regressive staining).

You should keep a personal record of the times you found most satisfactory to give you some guide for future work.

## 1.6. DIFFERENTIATION

If your tissue has been overstained, for example, the nucleus and cytoplasm are insufficiently distinguishable, you will need to remove some of the stain. The reagent used is known as a **differentiating agent** and the one most commonly used is **acid alcohol** (1·0 cm³ concentrated hydrochloric acid in 99 cm³ 70% alcohol).

After using acid alcohol wash the tissue in 70% alcohol.

## 1.7. DEHYDRATION

All traces of water must be removed from the tissues to allow complete penetration of the mounting medium.

### DEHYDRATING AGENTS

(1) **Industrial methylated spirit** (74° O.P.) is quite satisfactory as a dehydrating agent.

Dehydration must be effected gradually by transferring the object through alcohol of gradually increasing strength. The alcohols used could be 30%, 50%, 70%, 90% followed by 2 changes in undiluted industrial meths.

For school purposes undiluted industrial meths may be used instead of absolute alcohol. Absolute alcohol is 100% ethyl alcohol and is too expensive for general school use.

Dehydration, particularly the last stages, should be carried out in covered watch glasses to prevent dilution of the alcohols with water from the atmosphere.

*Preparation*—Use the following formula to calculate the volume of water which needs to be added to make up the various strengths of alcohol.

To prepare an '$x$% solution' from a '$y$% solution':
  Dilute $x$ cm³ of the $y$% solution to $y$ cm³ with water.
e.g. To prepare a 50% solution from a 70% solution:
  Dilute 50 cm³ of the 70% solution to 70 cm³ with 20 cm³ of water.

Start dehydration using an alcohol of slightly higher strength than the solvent of the stain.

*Dehydration must be complete.* If there is any water left in the tissues it will be apparent immediately the tissue is transferred to the clearing agent, the object then appears cloudy.

(2) **'Ethex'** ('Cellusolve')—Ethex is the commercial name for ethylene glycol monoethyl ether. This substance is miscible with water, alcohol and xylene.

It may be used as a dehydrating agent for *thin* sections and *small* whole mounts. It does not need to be diluted and tissues may be placed straight in it for dehydration after they have been in an aqueous or alcoholic stain.

## 1.8. CLEARING

As the name suggests one of the purposes of this stage is to make the stained tissues transparent enough for detailed microscopic examination. When alcohol or 'Ethex' has been used as the dehydrating agent all traces must be removed before the object is mounted in a medium with a higher refractive index, for example, Canada balsam or Micrex.

Liquids used to remove dehydrating agents are known as clearing agents. A commonly used clearing agent is xylene. There are two disadvantages with its use, it tends to make tissues brittle and if the object has been incompletely dehydrated an emulsion is formed between water and xylene. If you use xylene leave the object immersed for the shortest time necessary to effect clearing.

Benzene, toluene, terpineol and cedarwood oil may also be used as clearing agents.

## 1.9. MOUNTING

During this stage the object is placed in a mountant on a glass slide and covered with a coverslip.

## MOUNTANTS

### (1) Canada balsam in xylene

The disadvantages with using this mountant are that with exposure to light it turns a darker yellow, and that with time stock solutions may thicken and need to be diluted with the clearing agent before use. Apart from these disadvantages it is a useful mountant.

(2) **Micrex**

This mountant is colourless and does not colour on keeping. It seems to remain easier to handle than Canada balsam.

(3) **Euparal**

When euparal is used as the mountant a separate clearing agent is not necessary after dehydration using alcohol, the object may be transferred straight to this mountant after being immersed in alcohol.

METHOD OF MOUNTING AND COVERING

*Glass slides and cover slips must be clean.*

You will find the circular cover slips easier to use than the square ones.

You will almost certainly need practice before you are able to judge the correct amount of mountant to use.

A drop of the mountant should be placed in the centre of the glass slide. The object should then be put on top of the centre of the drop of mountant and the cover slip should be *gently lowered* with the aid of a mounted needle on to the mountant and object.

1.10. LABELLING

An adhesive label should be placed at one end of every slide which you intend to keep. Gummed slide labels with lines are obtainable in packets from biological suppliers.

*The label should carry the following information*

Name and brief description of object, e.g. *Lumbricus* T.S. intestinal region
Name of fixative and stain
Name of maker
Date of preparation

## 2. A SIMPLE STAINING TECHNIQUE FOR THE DIFFERENTIATION OF THE NUCLEUS AND CYTOPLASM

The nucleus and cytoplasm of the leucocytes in human blood are clearly distinguishable after treating a blood smear with Leishman's stain.

*Materials required*—Disposable sterile lancet—obtainable from Philip Harris or T. Gerrard
2 microscope slides—suitable size 75 mm × 25 mm
1 rectangular cover slip—size to fit above slide 50 mm × 24 mm
Microscope
Methylated spirit
Distilled water
Micrex or Canada balsam
Leishman's stain

*To prepare a stock solution of Leishman's stain*—
Leishman's stain 0·15 g
Methyl alcohol 100% 100 cm³
(neutral and free from acetone)

*Procedure*

(1) Obtain a drop of blood. Clean the underside of one finger tip with methylated spirit. Use a sterile lancet to pierce the skin and obtain a drop of blood. Place a *small* drop of blood near the centre of one of the *clean* glass slides.
(2) Convert the drop of blood into a smear. Take the second slide and hold it in an inclined position at 45° with the short edge touching the other slide just behind the drop of blood. Pull the inclined slide firmly and evenly through the drop of blood making a thin smear.
(3) Wave the slide in air to dry the smear.
(4) Cover the smear with a few drops of Leishman's stain and leave for 1 min.
(5) Add a few drops of distilled water and gently tilt the slide from side to side to mix the water and stain. Leave for about 10 min.
(6) With a filter paper held at the side of the slide draw off the stain.
(7) Pipette on distilled water to wash the surface of the smear.

(8) Dry the smear by placing the slide on the hood of a bench lamp for a few minutes.

(9) Place a drop of mountant (either Micrex or Canada balsam) on the smear and carefully lower a cover slip on to the slide.

(10) Examine under the low power and high power objectives of your microscope.

(11) Make drawings of some of the different types of cell you can see.

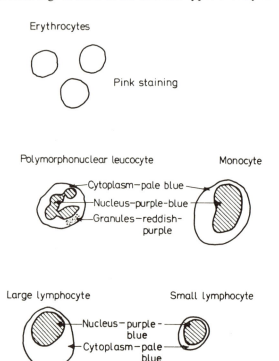

Fig. C3. *Diagram to show the colours of blood cells after being treated with Leishman's stain*

## Further work

Make a blood smear using blood from a frog or any amphibian available to you. Compare the appearance of the stained cells with those in the smear of human blood.

## 3.  A PERMANENT PREPARATION OF TEASED, STRIATED MUSCLE FIBRES—a method involving COUNTER STAINING

*Materials required*—Striated muscle (skeletal muscle from a dead laboratory animal)
Dissecting instruments
Microscope slide and cover slip
Microscope
Covered watch glasses
Ehrlich's haematoxylin
Eosin in 90% alcohol
70% alcohol
90% alcohol
'Absolute alcohol' (undiluted industrial methylated spirit)
Acid alcohol:
    99 cm³ 70% alcohol
        1 cm³ concentrated hydrochloric acid
Ammoniated 70% alcohol:
    99·5 cm³ 70% alcohol
        0·5 cm³ ammonium hydroxide
                        Sp. gr. 0·880
Xylene
Micrex

*Procedure*

Remove a small piece of striated muscle.
Place in a watch glass containing 70% alcohol as the fixative.
Tease the muscle gently but thoroughly to separate the fibres.
The best way of handling the few muscle fibres required is to transfer them on a section lifter.
Stain in Ehrlich's haematoxylin for at least 5 min.
*Differentiate in acid alcohol.
Wash in 70% alcohol.
*'Blue' in ammoniated alcohol.
Transfer to 70% alcohol.
Transfer to 90% alcohol.
Stain in eosin in 90% alcohol for 2 min.
*Wash in 90% alcohol.
Transfer through 2 changes of absolute alcohol.
Clear in xylene.

Mount in Micrex.

Cover and label.

*At these points you should examine the fibres under a micro-scope to make sure that you have achieved the required result before proceeding to the next stage.

The nuclei and bands of the striated muscle fibres should be stained purple blue with the haematoxylin. The cytoplasm will be stained pink by the eosin.

## 4. EXAMINATION OF LOCUST TESTES TO SHOW THE NUCLEI DURING MEIOSIS

During the formation of mature spermatozoa in the testes of the locust, the nuclei of the cells forming sperm go through a series of changes known as **meiosis.** Meiosis may be divided into phases, each phase being recognisable by the appearance and arrangement of the chromosomes from the nucleus of each cell.

In the male locust the diploid number of chromosomes is 23, 11 pairs of autosomes and a single sex chromosome X. As a result of meiosis 2 types of haploid spermatozoa are formed, one type has 11 autosomes and an X chromosome, the other type is without the X chromosome but has the 11 autosomes.

In order to see as many as possible of the nuclei of the testes cells in a state of meiotic division it is probably best to use adult male locusts a few days after the last moult.

*Materials required*—1 male locust (*Locusta migratoria* and *Schisto-cerca gregaria* are both suitable)

Killing bottle

Chloroform

Ether

Dissecting instruments

Pins

Wax bottomed dissecting dish

Microscope

Watch glass

Slides and cover slips

2 Petri dishes

Insect saline—Anhydrous calcium chloride 0·02 g, sodium chloride 0·7 g, distilled water 100 cm³

Aceto-orcein—Boil 100 cm³ of 45% acetic acid. Add 1 g orcein and boil gently for 30

min. Allow to cool. Shake well and filter.
90% alcohol
'Absolute alcohol' (undiluted industrial methy-
    lated spirit)
Euparal
Bunsen burner or spirit lamp

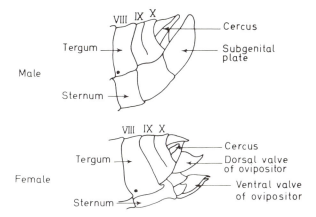

Fig. C4. *Lateral views of posterior end of abdomen of male and female* Locusta migratoria migratorioides

*Method*

Kill the locust by exposure to vapour from 50:50 mixture of chloroform and ether in killing bottle for at least 15 min.

Pin the male locust out dorsal side uppermost in a wax bottomed dish.

Make a cut in the mid dorsal line to open up the abdomen.

Pipette on a *few drops* of insect saline to prevent drying of the internal organs *(see Fig. C5)*.

Remove the mass of tissue consisting of the testes and fat material and place it in a watch glass containing a few drops of insect saline. (If you wish to preserve the follicles of the testes it is best to store them in acetic-alcohol (Clarke's fluid) as the fixative.)

Separate out a few of the follicles from the surrounding fat and place them on a clean microscope slide.

Gently lower a second glass slide down on to the follicles squash-ing the material to make a smear on both slides. Separate the two slides.

Place a few drops of aceto-orcein on each smear—this acts as a fixative and stain.

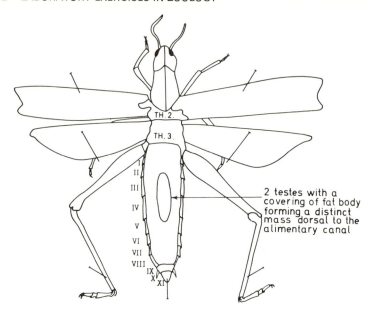

TH.2.

TH.3.

I
II
III
IV
V
VI
VII
VIII
IX
X
XI

2 testes with a
covering of fat body
forming a distinct
mass dorsal to the
alimentary canal

Fig. C5. *Diagram to show the position of testes in the abdomen of the male locust (dorsal view after abdomen has been opened by removal of the terga)*

(1)  Treat one slide as a temporary preparation.

Place a cover slip on the stained smear.

Warm the slide *very gently* for a few seconds.

Examine the smear under low power and try to distinguish separate cells. Select a suitable part of the smear for examination under high power. You should be able to see the darker staining chromosomes of the nuclei clearly distinguishable from the lighter staining cytoplasm of the cells.

(2)  If you wish to make the stained smear on the second slide into a permanent preparation treat it as follows:—

Place the slide in a covered Petri dish containing 90% alcohol.
.Transfer to a second covered Petri dish containing 'absolute alcohol'.
Mount in Euparal.

Make large drawings to show the appearance of some of the cells in your smear. Use textbook diagrams, photographs and any permanent preparations available to help you to identify the phases of meiosis reached by the cells you have drawn.

Compare your results with those obtained by other members of the class. You should be able to see cells in several phases of meosis.

## PART D

# Methods of Management and Culture of some Animals in the Laboratory

Before attempting to keep any animal in captivity it is important to ensure that one is able to provide those conditions which will enable the animal to remain healthy and possibly to reproduce. This means providing a suitable container for each species, a regular supply of food and water and in some cases fresh bedding material. Before any animal is accepted by a school, a member of the teaching staff, a laboratory technician, or a pupil must be found who will accept the day to day responsibility of feeding, watering and cleaning. Many pupils enjoy looking after animals and providing that they are reliable individuals there is no reason why animals should not be entrusted to their care with some supervision from a member of staff who should be ready to accept the ultimate responsibility for the well-being of the animal.

Remember that captive animals still need to be cared for during the weekends and in school holidays.

Everyone in charge of an animal should be taught how to handle the animal with the minimum of discomfort to it.

Respect for animals in captivity should be encouraged at all times.

## 1. GENERAL REQUIREMENTS FOR A SCHOOL ANIMAL ROOM

Many schools do not have the advantage of a separate room to house animals but are able to set aside part of the biology laboratory for the purpose of accommodating at least a few living creatures. The requirements listed below are relevant both for the laboratory and the separate animal room.

### STAGING

Where space is limited it is important to make the best use of what there is available. Cages and aquaria are best kept above ground level on benches or shelves. A number of cages may be held on adjustable shelving suspended from a wall, it is an advantage to be able to alter the position of the shelves as otherwise the height of the cages or aquaria is limited. The lowest shelf should be mounted about 1 ft from the floor, to enable easy cleaning, and the highest should not be over 5 ft above floor level. Additional shelves may be arranged between these two according to the height of cage to be accommodated. Solid shelves rather than slats will prevent the lower shelves being dirtied by the contents of cages above.

### SINKS

There should be at least one large sink with a hot as well as a cold water supply. If possible each heavy aquarium should be placed near a sink to make easier the process of emptying, cleaning and re-filling.

### WINDOWS

When windows are used to ventilate the room care must be taken to see that none of the cages is exposed to draughts. Over ventilation should be avoided in the winter months. Cages and aquaria should not be sited so that they are in direct sunlight. Open windows should be fly proof.

### HEATING

Some methods of heating the room will be necessary during the colder months of the year. If it is possible to maintain the air temperature of the room, $60°F$ ($16°C$) is the temperature which seems to suit most species of animal kept in schools. If tropical aquaria are to be kept fused power plugs should be available.

STORAGE

Containers will be needed for the following purposes:—
(1) To house the animals.

Suitable cages for various species are described later in the relevant sections. Containers with animals in should be labelled with the name, the sex, and if known the date of birth of the animal. It is also helpful if brief instructions for feeding, watering and cleaning are included on the label. If a pupil is in charge of the animal then his/her name should also be on the label.
(2) To store food material.

All food should be stored in covered containers inaccessible to pests.
(3) To store bedding material and aquarium sand.

Covered tea chests are useful for storing sawdust, paper or wood wool or torn newspaper. Sand may be stored in buckets or similar containers.
(4) To hold soiled litter and droppings.

Polythene buckets with tightly fitted lids are suitable for this purpose. These buckets should be emptied daily.


Miscellaneous equipment which will need storage space:—
(1) Cleaning equipment.

Detergents. Disinfectants. Cloths and brushes. Rubber tubing of about 1·25 cm bore for siphoning water from aquaria.
(2) Equipment for handling and transporting animals.

Thick rubber gloves. Nets for fish and amphibia.
(3) Killing apparatus.


## 2. MANAGEMENT OF THE STOCK

2.1. RECORDS

When written records are kept it is important that the facts recorded are accurate and that the information is in a form which may be easily understood by others referring to the records at a later date.

Records may be kept in a loose leaf book or file so that additional information may be inserted easily.

When an animal is purchased, the date of purchase and the supplier from whom it was bought should be recorded.

The animal should be given a name or number and this should be used in all written descriptions of the animal.

If the animal is to be used for breeding, records should be made to provide the following information:—the names or reference numbers of the two mated animals, the date of mating, the date of birth of the offspring and where possible the number of each sex produced. Any of the offspring which are to be kept as stock should be named or numbered.

Breeding records may be kept on cards, each breeding female having a card or cards to herself (*see Part E4*).

## 2.2   PREVENTION OF CROSS INFECTION

Any stock which is brought into the laboratory should be kept in a separate container from the resident stock for a short quarantine period e.g. two weeks.

Animals should only be put into clean cages.

A dirty cage should be thoroughly cleaned with a solution of soap or detergent, rinsed with clean water and dried. Polypropylene cages may be autoclaved up to about 120°C.

Cages should be disinfected at regular intervals and always after they have been used to house a sick animal. Suitable disinfectants are:—

The hypochlorites, e.g. Parazone, Chloros and Domestos.

Dettol.

1% solution of Cetrimide B.P., Cetavlon, or Tego M.H.G.

Do not use Cresol and strong phenolic disinfectants.

The possibility should be borne in mind that disease-causing organisms active in a laboratory animal may also cause disease in man. After handling any animal the hands should be washed thoroughly before other objects are touched.

Wounds received as a result of bites or scratches should be washed immediately with soap and water. A mild disinfectant should then be applied. A wound which fails to heal in the normal time or whose surrounding tissues become swollen or very inflamed should receive expert medical attention.

## 2.3.   BREEDING

Sexually mature animals of different sex should only be placed together if young stock are required for some definite purpose. Make sure beforehand that you have enough room and resources to keep the young animals.

Listed below are some of the reasons for breeding animals in schools.

(1) To maintain or increase the school stock.

(2) To provide animals for sale, the proceeds helping towards the cost of upkeep of the school stock.

(3) To provide subjects for experiment with the object of learning more about the physiology of the animal.

(4) To provide material for dissection so that the anatomy of the animal may be investigated.

(5) To find out something about the inheritance of characteristics.

## 2.4. ANAESTHESIA FOR COLD BLOODED AQUATIC ANIMALS

It may be desirable to immobilise temporarily an animal if it is to be examined without difficulty.

Cold blooded aquatic animals may be anaesthetised in a solution of MS–222 Sandoz. This anaesthetic allows complete recovery.

It is a white, crystalline powder obtainable directly from:—

Sandoz Ltd.,
Sandoz House,
23, Great Castle Street,
London, W.1.

T. Gerrard & Co. Ltd.,
Gerrard House,
Worthing Road,
East Preston,
Near Littlehampton,
Sussex.

Philip Harris Ltd.,
63, Ludgate Hill,
Birmingham, 3.

MS–222 Sandoz is very soluble in water and will readily dissolve in sea water as well as fresh water. It forms a clear, colourless acid solution.

Solutions of concentration 1 g:1,000 cm³ water to 1 g:30,000 cm³ may be used. You will need to try various concentrations to see which is best for your requirements. The time taken to anaesthetise an animal depends on the concentration of the solution, the temperature of the solution and the weight of the animal.

A solution which seems to be effective for most small fish and amphibians has a concentration of 1 g:3,000 cm³ water.

Solutions should be freshly prepared before use.

The animal may be anaesthetised by immersion in the solution. Air-breathing cold-blooded animals should have the head supported above the level of the liquid during anaesthetisation. Once anaesthetised the animal should be removed from the solution.

## 2.5. HUMANE KILLING

When killing an animal is necessary it should always be carried out by a member of staff. Killing is most humane when a state of unconsciousness is reached rapidly. Unconsciousness may be produced in two main ways, either by mechanical damage to the central nervous system or by exposure to a chemical which anaesthetises the animal. Animals should never be drowned as this is a slow method of killing and may cause unnecessary suffering.

PHYSICAL METHODS OF KILLING  (Recommended by the Universities Federation for Animal Welfare)

*Small fish*—Fish weighing less than 250 g are best killed by holding the fish in the left hand, placing the right thumb in the mouth and the right fore-finger at the junction of the head and neck, and bending the head sharply until a crack is felt.

*Reptiles and amphibia*—These should be killed by a sharp blow on the head cr by decapitation with bone forceps, secateurs or a sharp pair of scissors.

*Mice*—The mouse is held by the tail and placed on a surface that it can grip, when it will stretch itself out so that a pencil or similar object can be placed firmly across the back of the neck. A sharp pull on the tail will then dislocate the neck and kill the mouse immediately.
Or the animal may be held as before and hit hard behind the ears with a blunt instrument.

*Rats*—If the animal can be handled, place it on a duster and wrap the duster firmly round the body, including both front legs; kill by concussion thus:—
(1) Hold the head downwards and strike very hard behind the ears with a stout wooden stick.
(2) Hold firmly, belly upwards, and strike the back of the head *very hard* against a hard horizontal surface.

It is appreciated that some school staff will find these physical methods of killing animals unpleasant and a chemical method recommended by The Universities Federation for Animal Welfare is included below.

## CHEMICAL METHODS OF KILLING

(1) **Carbon dioxide**—The Universities Federation for Animal Welfare recommends carbon dioxide as the most suitable chemical for the humane killing of small animals.

*Necessary Equipment for carbon dioxide euthanasia using a plastic bag*

Cylinder of carbon dioxide.
   This may be obtained from:—

| | |
|---|---|
| Distillers Chemical and Plastics Ltd., | The British Oxygen Co. Ltd., |
| Devonshire House, | P.O. Box 12, |
| Mayfair Place, | Walkden, |
| Piccadilly, | Manchester. |
| London, W.1. | |

   A short length of rubber or plastic tubing to connect the cylinder with the container holding the animal.
   A wire cage (or some similar container). Cages may be obtained from:—

> Associated Crates Ltd.,
> 255, Gray's Inn Road,
> London, W.C.1.

   Transparent polythene bags of approximately five times the volume of the wire cage.
   Rubber bands to close the mouth of the bag.

*Technique*

The animal is placed in the wire cage and is allowed time to settle down, the cage is then placed inside a polythene bag. The bag is pressed close to the sides of the cage and the end of the tubing from the carbon dioxide cylinder is introduced through the mouth of the polythene bag. The mouth of the bag should be held tightly with one hand around the tubing while the other hand turns the cylinder key to release a slow flow of gas. The bag should be inflated until it is full of gas but not under pressure. The animal should become unconscious when the bag is half full of gas. The tubing should be withdrawn when the bag is full, the gas turned off and the mouth of the bag closed with a rubber band. The animal should be left in contact with the gas for a further 10 min after all signs of breathing have ceased. The bag may then be opened and the dead animal taken from the cage.

Plate D1. *Captive rat in a wire cage inside a polythene bag (From* Humane Killing of Animals *by courtesy of U.F.A.W.)*

In the case of mice, rats, hamsters, rabbits, budgerigars, pigeons and chickens, several animals of the same species may be placed in the same cage provided there is sufficient floor space and the animals do not fight.

(2) **The use of chloroform** (Chloroform should never be used to kill rabbits)

*Necessary equipment*—Desiccator (internal diameter at top 25 cm)
　　　　　　　　　　Chloroform
　　　　　　　　　　Cotton wool

*Technique*—A pad of cotton wool is soaked in chloroform and is placed under the gauze in a desiccator. The animal is placed in the upper half of the desiccator and is exposed to chloroform vapour. It should be left for 10 min after all signs of breathing have ceased.

## 3. CULTURAL METHODS

Methods of culture of only those animals which are relatively easily kept in schools are described.

*Addresses of Biological Suppliers from which all the listed animals may be obtained*

(1) Philip Harris Ltd.,
    63, Ludgate Hill,
    Birmingham, 3.

(2) T. Gerrard & Co. Ltd.,
    Gerrard House,
    Worthing Road,
    East Preston,
    Near Littlehampton,
    Sussex.

### 3.1. *Paramecium caudatum*

This is a useful member of the *Protozoa* to keep in schools. *Paramecium* may be used to demonstrate ciliary movement, the functioning of the contractile vacuole and some parts of the processes of ingestion and digestion. It illustrates a high degree of complexity for an acellular animal.

*Source*—Obtain *Paramecium* either by collection from fresh water or from one of the biological suppliers.

*Suitable containers*—Glass beakers, jars or small aquaria.

*Management of culture*—*Paramecium* will live and reproduce in a medium made in the following way. Pick 6 g of Timothy Grass (*Phleum pratense**) and cut the stems and leaves into 1 in long pieces. Boil the grass for 20 min in a flask containing 1 litre of water. Place a beaker on top of the flask and boil for a further 15 min. Allow to

*Other grasses may provide an equally good infusion.

cool and stand, still covered, for 24 h. A pH of approximately 7·8 is advisable for successful culture. Pour this infusion into the required container and add *Paramecium*.

## 3.2. *Turbellaria*

The fresh water turbellarians which are likely to be encountered during collecting are commonly described as planarians. Although this term includes a variety of species they may all be cultured in a similar way.

Planarians are useful subjects for behaviour experiments. They normally show good powers of regeneration and may be used to demonstrate this property.

*Source*—Examine the under surface of submerged stones, leaves, boards etc., for planarians. Where they are known to be present in flowing water they may be baited by small pieces of fresh liver suspended in the water, they may then be collected as they are attracted to the liver. Collect enough water at the same time to fill the laboratory containers which will be used to house the planarians.

*Suitable containers*—Glass or plastic containers with a relatively large air-water surface area compared with the volume of the container.

The water which was collected at the same time as the planarians should be boiled (to kill disease producing parasites) and cooled before it is used to fill the containers.

*Management of culture*—Planarians will die in contaminated water and they should be transferred to clean containers each week. They need feeding once a week and it is most convenient to transfer the worms to a clean container after they have been fed.

*Feeding*—A small piece of fresh beef liver should be given and this should be left in the container for about four hours after which all the remaining liver should be removed.

*To increase the stock*—Some of the stock worms should be cut transversely using a very thin cover slip. Remove the tail pieces to separate containers. After 4 weeks they should have regenerated anterior ends and feeding can then begin.

### 3.3. *Lumbricidae* (Earthworms)

In schools, earthworms are used chiefly for two purposes: for dissection and as a food for other laboratory animals but it is possible to learn something about their behaviour and physiology if a culture is kept in school.

*Source*—Worms may be collected from land near the laboratory or may be purchased from a supplier. When you are collecting it is probable that you will dig up more than one species of worm, if you only want to house one species make sure you separate the worms before introducing them into your culture.

*Containers*—G. J. Ashby in the UFAW Handbook recommends the following type of container and culture medium for keeping earthworms.

A wooden box made of boards 1·5 cm thick which are jointed and screwed together. The boards should be cut to give a box of the following dimensions, length 46 cm, width 38 cm and depth 20 cm. Angle irons may be used to give additional support. The wood should be treated with a wood preservative. Holes 3 mm in diameter (1 hole per square decimeter) should be drilled in the base of the box to allow drainage. To lift the box clear of the surface of the bench it should be supported on a wooden frame.

To prevent the worms escaping, a top for the box could be made from perforated zinc or wire gauze.

*Culture medium*—1 part manure, (horse (best) or rabbit)—collected and dried out and subsequently remoistened to give a water content of 30%
3 parts soil—light sandy soil with all the lumps removed
5 parts peat moss—treated with hydrated lime to give a pH 7·0 or 7·5
A small sprinkling of sharp sand

The soil, peat moss and sand should be mixed together and moistened to give a water content of approximately 30%. This mixture should be used to fill the box to about 6 cm from the top and the dung placed in a layer on the top of this. The worms may then be placed on the surface and allowed to burrow down into the box.

*Management of culture*—The contents of the box should be sprinkled with water daily to keep the water content of the medium around 30%.

When the amount of food appears to be getting short dung should be spread on the surface of the culture.

Once every two or three weeks the culture should be tipped out on to a tray. This will enable you to examine the worms and collect any cocoons, and in repacking the culture medium you will also be helping to aerate it.

The culture medium should be completely changed every 6 months.

Any cocoons collected could be placed in freshly prepared culture boxes to start off new cultures.

### 3.4. *Daphnia*

This animal has a transparent exoskeleton and thus may be used to demonstrate some of the features of crustacean anatomy. The heart beat can be clearly seen under a microscope and use may be made of this fact in constructing simple physiological experiments.

These animals are also useful as food for cultures of *Hydra*, fish and newly metamorphosed *Xenopus*.

*Source*—Obtain by collection locally from small ponds or lakes or from one of the suppliers.

*Containers*—It is not always an easy matter to maintain a culture of *Daphnia* in a school laboratory and the type of container seems to have an effect on the life-span of the culture. Porcelain sinks and large wooden tubs are recommended as giving satisfactory results. The position of the container is probably also important and too much light should be avoided. Cultures in small glass containers can be satisfactorily maintained for short periods.

*Feeding*—A suspension of yeast may be used as food. A few drops should be added at intervals of a few days, care must be taken to avoid fouling the water by overfeeding.

*Aeration*—Sufficient oxygen to aerate the water may be produced by growing *Elodea canadensis* in the container.

*Sub-culture*—Some of the *Daphnia* should be moved to a fresh culture container every few months.

## 3.5. *Periplaneta americana*

These animals may be kept in schools for a variety of reasons. They can be used as the subjects in behaviour experiments (choice chambers, and experiments designed to show circadian rhythm), their method of locomotion may be observed and they show a life history with only a slight metamorphosis. They may also be used for dissection but it may well be cheaper to buy preserved specimens for this purpose unless your school stock contains enough cockroaches near maturity to replace those killed.

*Source*—Live cockroaches may be obtained from one of the suppliers listed above (*see* beginning of section 3).

*Container*—A small glass or plastic aquarium with a loosely fitting glass top acts as a suitable container. An inch of sawdust should be placed in the bottom of the container and this need not be disturbed for six months to a year. Balls of newspaper and pieces of corrugated cardboard should be placed to one end of the container and water and food containers at the other end.

*Management*—Provided that a temperature of 25°C and a reasonably humid atmosphere can be maintained these insects are quite easy to keep in the laboratory. If the air temperature of the laboratory or the animal room cannot be kept at 25°C the container should be heated with an electric light bulb. For breeding a slightly higher temperature of around 27–28°C is preferable. A conical flask should be three-quarters filled with water and a roll of filter paper, to act as a wick, should be inserted through the neck of the flask. This will provide a humid atmosphere and a source of water for the cockroaches to drink.

The food container should be filled with bran or porage oats.

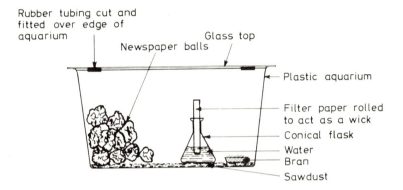

Fig. D1. *Diagram of cockroach container*

Food, water and clean balls of newspaper should be given when necessary. When the contents of the container are being replenished the insects do not usually attempt to escape but scurry to hide under the newspaper. As cockroaches are potential pests care should be taken to ensure that they do not escape.

The culture should be inspected regularly for egg capsules which may be collected and put to develop in another container. The first

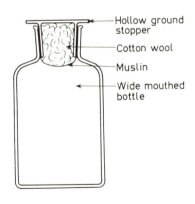

Fig. D2. *Diagram of bottle suitable for killing cockroach*

instar is very susceptible to drying and should be kept at 70% RH for the first week. The older stages will live quite well in lower humidities providing they have access to water.

*Handling*—The cockroach should be firmly gripped round the middle abdomen with a long pair of fine forceps.

*Killing*—These insects are best killed by chloroform vapour in a killing bottle (*see Fig. D2*).

Remove the stopper and apply a few drops of chloroform to the cotton wool in the stopper. (If a small piece of butter muslin is secured round the stopper the cotton wool will be prevented from falling out.) Place the cockroach inside the bottle and replace the stopper.

### 3.6. *Locusta migratoria migratorioides*

This species of locust may be kept and bred successfully in captivity providing that daily attention is given to the culture.

It is a large enough insect to be used for dissection to demonstrate features of insect anatomy. The testes provide good cells for a study of meiosis (*see Part C4*). It is possible to observe a complete life cycle during one school term. At 28°C the incubation period of the eggs is about 16 days, this is followed by about one month of nymphal life and the adults mature 4 weeks after the wings have fully developed. At 34°C the incubation period is 11 days and the length of nymphal life is 26 days.

*Source*—Living locusts may be obtained from both the suppliers listed (*see* beginning of Section 3).

Plate D2. *Position of male and female locusts during pairing*

*Distinguishing characteristics of the male and female locust*

| Mature male | Mature female |
| --- | --- |
| (1) Exoskeleton bright yellow | (1) Exoskeleton dark brown |
| (2) Posterior abdomen ends ventrally in an upturned keel shaped structure called the sub-genital plate (*see Fig. C4*) | (2) Dorsal and vental valves of the ovipositor are visible at the posterior end of the abdomen (*see Fig. C4*) |
| (3) The cercus is relatively longer (*see Plate D3*) | (3) The cercus is relatively shorter (*see Plate D3*) |

*Container*—The cage illustrated is basically the type of cage described by Philip Hunter-Jones in 'Rearing and Breeding Locusts

Plate D3. *Lateral view of posterior end of abdomen of male (right) and female (left)* Locusta migratoria migratorioides

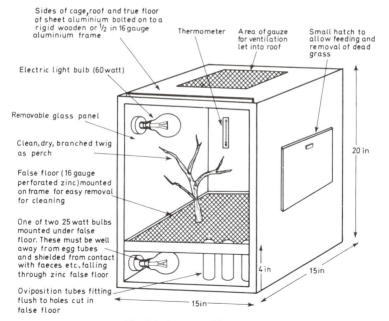

Sides of cage, roof and true floor of sheet aluminium bolted on to a rigid wooden or ½ in 16 gauge aluminium frame

Thermometer

Area of gauze for ventilation let into roof

Small hatch to allow feeding and removal of dead grass

Electric light bulb (60 watt)

Removable glass panel

Clean, dry, branched twig as perch

False floor (16 gauge perforated zinc) mounted on frame for easy removal for cleaning

One of two 25 watt bulbs mounted under false floor. These must be well away from egg tubes and shielded from contact with faeces etc., falling through zinc false floor

Oviposition tubes fitting flush to holes cut in false floor

20 in

4 in

15 in

15 in

Fig. D3. *Diagram of locust cage*

in the Laboratory', published by the Anti-Locust Research Centre.
Cages are obtainable from:—T. Gerrard & Co. Ltd.
Philip Harris Ltd.
Griffin and George

Details are given of the construction of the cage which may be
made fairly cheaply in a school workshop.

*Preparation of the sand filled tubes for oviposition*—If mature locusts
are kept for breeding purposes it will be necessary to provide the
females with sand in which they can deposit their egg pods. Suitable
sand containers, as used by the Anti-Locust Research Centre, are
aluminium tubes of 10 cm depth and 3·75 cm diameter. These tubes
are obtainable from Philip Harris Ltd.

The false floor at the front of the cage should be drilled with three
or more holes so that when an aluminium tube is placed in each hole
the base of the tube rests on the true floor and the opening of the
tube is flush with the false floor.

Each tube must be filled with moist sand prepared by mixing 15
parts by volume of distilled water to 100 parts by volume of clean,
dry, sterilised, coarse sharp sand. Weigh the tube of damp sand and
until eggs have been laid reweigh the tube daily, if necessary adding
distilled water to maintain the correct weight.

MAINTENANCE

*Temperature*—A suitable daytime temperature is 34°C, this may be
achieved by switching on all or some of the electric light bulbs of the
recommended wattage (*see Fig. D3*) fixed inside the cage.

A lower night temperature of about 28°C is tolerated and this may
be reached without the use of the 60 W bulb.

Locusts may become sterile due to the heat if the air temperature
of the cage rises above the recommended temperature.

*Humidity*—If *just* sufficient fresh grass is given daily, it is probable
that the water content of the cage will not be too high. Signs that the
humidity is too high are soft, wet faeces and/or condensation on the
glass front of the cage.

The interior of the cage should be kept as dry as possible. Bacterial
and fungal disease may occur if the humidity is too high.

*Feeding*—Fresh grass, cut into lengths and made into bundles should be placed in a small water-filled jar inside the cage every day. It is important to give enough grass, as if insufficient is given the locusts will turn cannibalistic.

It is a good plan to change the source of grass occasionally as one source may be contaminated with the eggs of a parasitic worm, the nematode, *Mermis*. This worm may reach a length of 12 cm in the body cavity of the locust and may eventually cause the death of the locust.

It is not necessary to provide a separate container for water when grass is given in water filled pots.

At the weekend when it may not be possible to provide fresh grass, a Petri dish of wheatbran and another of water-soaked cotton wool should be left in the cage.

*Daily care*—Uneaten grass, faeces, dead locusts or cast cuticles should be removed daily.

Just sufficient fresh grass should be provided.

Check the temperature and in the evening switch off the 60 W bulb.

If the sand filled tubes are in place, check that the sand has not become too dry.

*Cleaning*—It is advisable to clean thoroughly the empty cage every 4–6 weeks. The Anti-Locust Research Centre recommends the following routine:—

(1) Remove all traces of food and faeces. Discard any twigs or perches.

(2) Thoroughly scrub all parts of the cage with very hot water and then scrub all parts with detergent.

(3) Disinfect the outside and inside of the cage. Allow to dry before use.

*Incubation of sand filled tubes containing egg pods*—The sand filled containers should be removed and replaced by new ones every few days, once egg laying has begun.

Each tube should be covered with a metal or glass disc or a cap of aluminium foil to prevent evaporation from the sand.

The tubes should be incubated at a temperature of 28°–32°C.

After about 10 days at this temperature the covers should be removed and if the tubes are not already in the cage in which the young locusts are to be reared they should be placed in position resting on the false floor.

*Special points for attention during the rearing of locusts*—The nymphs must be reared in a cage separate from the one housing the mature locusts.

The newly hatched nymphs are very active and are less than 1·25 cm long. Care must be taken to ensure that they do not escape through any gaps in a badly constructed cage.

Food must be accessible to the newly hatched nymphs as they begin to feed soon after hatching.

It is necessary to provide perches for the nymphs to cling to while moulting is taking place.

## 3.7. *Drosophila melanogaster*

*Drosophila melanogaster* is useful for genetic experiments. The insects may be used to demonstrate monohybrid inheritance, dihybrid inheritance and linkage. They may also be used for behaviour experiments, population counts and as an example of an insect life history showing complete metamorphosis. Chromosome preparations of the salivary glands may also be attempted.

Their short life history, easy culture and prolific rate of reproduction make them a most useful animal to keep in the school laboratory.

*Source*—Both biological suppliers listed at the beginning of this section. In addition:—Miss B. Barton,
21, Hillingdon Rise,
Sevenoaks,
Kent.

Cultures of *Drosophila* received from the dealers should be placed in an upright position at a temperature of about 25°C.

The cultures usually contain all stages of the life history.

*Distinguishing characteristics of the male and female*

| Male | Female |
|---|---|
| (1) Smaller size of whole body | (1) Larger size of whole body |
| (2) Smaller abdomen | (2) Larger abdomen |
| (3) Posterior end of abdomen blunt | (3) Posterior end of abdomen pointed |
| (4) Sex comb on first tarsal segment of foreleg | (4) No sex comb |
| (5) Complex external reproductive organs at tip of abdomen | (5) See diagram for details of posterior abdomen |

(A) Lateral view of part of the abdomen to show regions of pigmentation

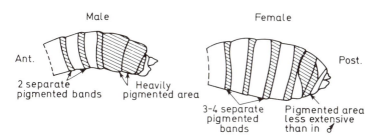

(B) Ventral view to show shape of abdomen

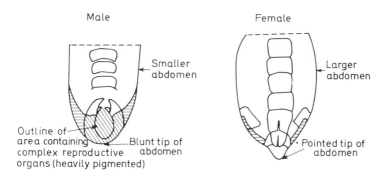

(C) Diagram of the left foreleg (similar to the right foreleg)
(for the sake of clarity most of the bristles have been omitted)

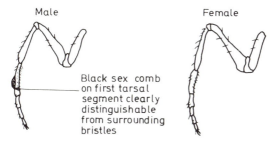

Fig. D4. *Outline diagrams to show the distinguishing features of the male and female* Drosophila

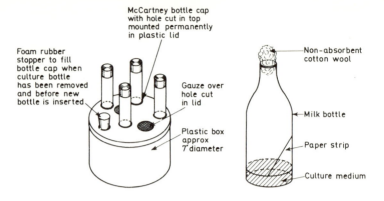

Fig. D5. *Main stock containers*

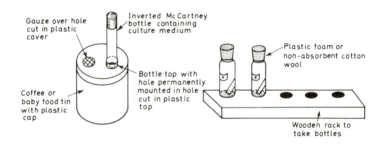

Fig. D6. *Sub-culture containers*

*Containers*

(1) Stock containers.
The most generally used are milk bottles, particularly one-third pint size, but plastic boxes and plastic lidded tins may be used equally well.
(2) Containers for small numbers of flies.
McCartney bottles or 2·5 cm diameter specimen tubes.

*Technique for subculturing in McCartney bottles from the main stock containers*—McCartney bottles from the main stock containers should be removed every 3 days or so and stood in racks ready for use. Replacement McCartney bottles containing culture medium should be inserted in the lid of the stock container.

*Culture media*—Most stockists offer a prepared culture medium but, it may be easier and less expensive to prepare your own, particularly if it is intended to keep culturing for some time. Culture media should be prepared in comparatively large quantities at one time and kept in stock bottles or tubes. These cultures can be re-sterilised before use. An anti-mould, e.g. Nipagin or an equivalent should be used in all media. Variations in traditional formulae should be tried particularly in the amount of agar-agar used. The adults and larvae feed mainly on yeast which is added prior to culturing.

(1) Agar-Agar   1 g
    Banana      One
    Water       250 cm$^3$
    Nipagin     0·4 g
    (Dried yeast made into a suspension and added when culturing)

The agar-agar should be completely dissolved in the water which should be warm. The banana should be mashed and added and the whole brought to the boil stirring continuously. When the mixture is boiling add the anti-mould. The mixture should simmer until the consistency is even, then pour it into a stock bottle or individual culture tubes or bottles. Use a funnel to introduce the culture medium into the tubes or bottles to avoid smearing the side of the containers.

(2) Maize meal 200 cm$^3$ (Very fine oatmeal can be substituted but should be soaked in some of the water for 24 h before use)
    Nipagin     0·5 g
    Treacle     75 cm$^3$
    Water       250 cm$^3$
    Agar-agar   5 g
    (Dried yeast made into a suspension with a little water and added when culturing)

Heat some of the water and dissolve the agar-agar. Add the rest of the water and continue to heat gradually adding the meal and treacle. Continue to heat until the mixture is fairly stiff then add the anti-mould. The medium should then be poured into containers as required and allowed to cool.

*Culturing*—Before placing the flies in the container cut a small strip of filter paper or paper towel (using forceps to hold the paper and strip) and after introducing a drop of yeast suspension on to the top of the culture medium place the strip in the container (*see Fig. D5*). If the tube is destined for the stock container the paper may be omitted until after the eggs have been deposited.

10–11 days is the length of the life cycle at 25°C (*see Part E3*).

*Mite infected cultures*—It is advisable to examine cultures frequently for the presence of mites. If present they may be seen on microscopical inspection of individual fruit flies. Cultures infected with mites should be destroyed.

*Obtaining virgin flies*—Virgin flies can be obtained by releasing all the adult flies from the required culture tubes. Those flies that appear during the following eight hours can be taken as virgin. The newly emerged flies will be seen to be lighter in colour than the older flies.

ANAESTHETISING PROCEDURE FOR THE EXAMINATION OF ADULT *DROSOPHILA*

*Materials required*—Culture tube containing adult flies
Empty corked tube of the same size to act as the anaesthetising tube
Short length of wire
Cotton wool
Ether
Fine tipped paint brush
White tile
Binocular microscope

N.B. Make use of the fact that adult flies are negatively geotactic. *See Fig. D7* for the anaesthetising procedure.

Having examined the anaesthetised flies and determined their sex those required should be placed in freshly prepared labelled culture tubes.

Do not allow the anaesthetised flies to fall on to the culture medium and become stuck on the surface. It is a good plan to remove the culture tube from the upright position, lie it on its side and with the aid of the paint brush introduce the flies into the tube so that while anaesthetised they remain resting against the side of the tube without coming into contact with the culture medium. The mouth of the culture tube should then be plugged with plastic foam or cotton wool and once the flies have begun to fly again the tube may be stood in the upright position.

## 3.8. *Tribolium confusum* (and *Tribolium castaneum*)

Both these species of flour beetle may be cultured in the same way. Both species may be obtained in a number of mutant forms showing variations in body and eye colour from those of the wild type.

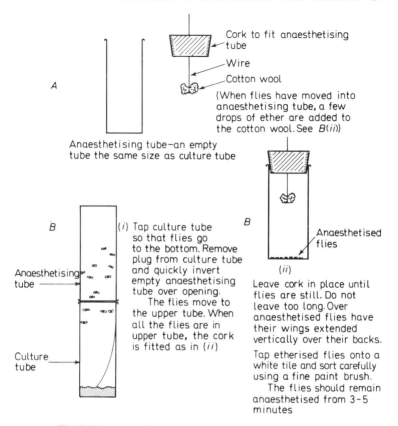

Fig. D7. *Diagrams of anaesthetising procedure for* Drosophila

They may be bred in the laboratory to demonstrate the genetic principles of segregation, dominance, recessiveness, independent assortment, autosomal and sex linkage.

Compared with *Drosophila* their use for genetic experiments has one disadvantage. The life cycle of *Tribolium* takes at least 6 weeks and thus 12 weeks are needed before an $F_2$ generation can be obtained.

*Source*—They may be obtained from both of the suppliers listed above (*see* beginning of section 3).

*Containers*—2 lb jam jars, kilner jars or other containers of similar size are suitable. The mouth of the jar may be covered with cotton

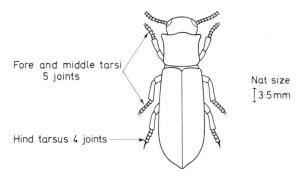

Fore and middle tarsi
5 joints

Nat size
⊺3·5mm

Hind tarsus 4 joints

Fig. D8. *Outline diagram of* Tribolium castaneum

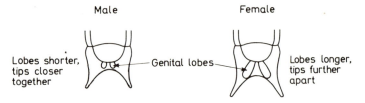

Male                           Female

Lobes shorter,
tips closer
together

——Genital lobes——

Lobes longer,
tips further
apart

Fig. D9. *Ventral view of the posterior end of the abdomen of* Tribolium *pupae*

cloth secured with a rubber band. The container should be two-thirds full of the food medium.

*Culture medium*—12 parts wholemeal flour to 1 part dried yeast. No water needed.

*Maintenance*—Unless pupae are required the cultures may be left undisturbed for 12 weeks. After this time it may be advisable to subculture. Rough handling of *Tribolium* impairs the fecundity of the females.

*Sexing*—The easiest time to distinguish between the sexes is when the beetles are in the pupal stage.
   You will need to use a binocular microscope of at least × 30 magnification.
   A pupa is oatmeal coloured and is 4–5 mm long.
   In the female pupae the external genitalia are more prominent than in the male pupae. *See Fig. D9* for differences in the posterior end of the abdomen of the pupae.

*Setting up a cross*—After the pupae have been sexed they may be segregated and crosses set up. Separating the beetles at this stage ensures that crosses are made with virgin females.

Pairs of pupae should be placed in small, covered specimen tubes (10 cm × 2·5 cm diameter) half filled with the culture medium. They require no maintenance until the $F_1$ emerges.

## 3.9. *Xenopus laevis*

Adult *Xenopus* are fully aquatic and may be kept successfully in laboratory conditions. As they can be induced to mate and produce fertile spawn by the injection of a gonadotrophin, a life cycle involving a mating process followed by external fertilisation, the development of larval forms and metamorphosis may be demonstrated.

The larvae are particularly useful for the observation of melanophores and the blood circulation.

Where there are sufficient stocks of adult *Xenopus* they may perhaps be used in preference to *Rana* as the amphibian type for dissection.

*Source*—Sexually mature toads may be obtained from one of the suppliers listed (*see* beginning of section 3).

*Distinguishing characteristics of the male and female*

(1) The size of the two adults is different. A mature female is much larger than a male and she is more pear shaped (*see Plate D4*).
(2) There is a difference in the cloacal region. The female has protruding labia which surround the cloaca while the male has very small labia which do not extent beyond the cloaca.

*Containers*—Unbreakable tanks of polyethylene are obtainable from the biological suppliers listed.

The tank should be filled with water to a depth of 12 cm and covered to prevent the toads escaping. A cover can be made with 2·5 cm mesh Claritex.

Plate D4. Xenopus, *showing relative size of female (left) and male (right)*

MANAGEMENT

*Temperature of water*—The toads will survive at normal room temperature so that a heating element in the tank need not be provided. The water temperature should not fall below 10°C.

*Feeding*—Small pieces of earthworm, chopped ox liver, spleen or heart will be acceptable. (*Xenopus* are completely carnivorous and will eat their own eggs and tadpoles.) Food should be given twice a week and care should be taken to see that food does not accumulate at the bottom of the tank.

*Cleaning*—The water in the tank should be changed once a week. The toads should be transferred to a reserve tank while their tank is being cleaned.

*Handling*—The toads should be caught in an aquarium net and one hand should be placed over the opening of the net to prevent the toads escaping. The toads are very difficult to catch by hand as their skins are extremely slippery. (For handling prior to and during hormone injection see below.)

*Diseases*—'Red-leg' is an infectious disease which may destroy a complete stock of toads. It is caused by a bacillus *Aeromonas hydrophila* (previously called *Proteus hydrophilus*). A toad suffering from the disease will have red patches caused by haemorrhages on the legs and webs. Any toads with these symptoms should be killed. Other toads which have been in the same tank should be removed and placed in separate containers in a dilute solution of mercurochrome. For each litre of water add 1 cm³ of 0·2% mercurochrome. (A 0·2% stock solution of mercurochrome contains 5 g in 2·5 litres.)

They should be examined daily for signs of disease.

Any new stock should be kept in a quarantine tank for a week or two and closely watched to see if they develop any signs of the disease.

BREEDING

*Timing the injections*—If you inject the toads at the beginning of an average school term the resulting tadpoles will be available for examination during the term and they may have completed their development by the end of term. Metamorphosis may occur 8–12 weeks after fertilisation when the tadpoles are kept at 23°C. However, it is possible that it will take your tadpoles longer than this to metamorphose as their development also depends on the amount of food and space which has been available to them.

*Extra equipment needed for breeding*

2 small tanks (30 × 20 × 20 cm), one with a cover. The tanks should be three-quarters filled with water preferably maintained at a temperature of 23°C. One tank, the spawning tank, could be provided with a false floor of 'netlon' slightly raised above the floor of the tank. The eggs as they are laid should then fall through the 'netlon' out of the reach of the adult *Xenopus* who otherwise might eat the spawn when they come out of amplexus.

Two 2 cm³ disposable syringes, each with a No. 18 hypodermic needle.

Ampoules of Pregnyl
 (chorionic gonadotrophin)
Ampoules of distilled water
⎰ These may be obtained from one of
⎪ the biological suppliers listed. Order
⎨ the 'medical pack' of Pregnyl, not the
⎪ veterinary pack containing phenol
⎩ which is lethal to toads.

Nettle powder. Obtainable from the biological suppliers.
Clean dry cloths.

### INJECTION TECHNIQUE

Injections should only be carried out by a teacher.

Only those toads which are mature and fully acclimatised to the school laboratory conditions should be used for injection.

The tank containing the Xenopus should be situated in a place where the toads are not disturbed by the movements of passing pupils. It is a good plan to inject the toads at the end of the afternoon and leave them undisturbed overnight.

In the laboratory a female toad may be stimulated to lay eggs and a male toad to shed spermatozoa after injections of the hormone chorionic gonadotrophin. Two injections are given to each toad; the first injection is given to each toad four days before the second injection. The purpose of the first injection is to bring the toad into breeding condition without the discharge of gametes.

*Preparation of the two syringes for the primer injection*

Female
(1) Break the top off an ampoule containing 1 cm³ distilled water and insert the needle of a syringe, draw all the water up into the syringe.
(2) Break the top off an ampoule containing 100 units of Pregnyl, insert the needle of the water filled syringe and expel all the water.
(3) Draw up the solution of Pregnyl into the syringe being very careful to expel all air bubbles and to withdraw all the fluid from the ampoule.

Male
Follow the same routine except in this case use 0·5 cm³ of the solution of Pregnyl.

*Handling Xenopus for the injection of Pregnyl*

Although it is quite possible for one person alone to hold the toad firmly with one hand and inject it with a syringe held in the other hand, we suggest that at least for the first few times you carry out the injection routine, two adults are present. At first it may be easier for one person to hold the toad while the other person does the injection.

The hormone solution is to be injected into a dorsal lymph sac situated just to the side of the mid line. A line of 'stitch marks' indicates the position of the dorsal septum.

The toad should be held very firmly in a clean dry cloth. Use the thumb to hold the hind leg in an extended position under the cloth. Insertion of the needle is easier if the skin of the toad is stretched. You will have more control over the toad if you rest your arm and the hand holding the toad on the bench. With the syringe in the other hand direct the needle through the skin of the thigh (*see Fig. D10*). Push the needle gently forwards through the septum and into the dorsal lymph space. The needle should be kept just below the skin so that there is no damage to internal organs. Expel the contents of the syringe. Withdraw the needle carefully. Place the toad in one of the small tanks. Inject the second toad and place it in the same tank.

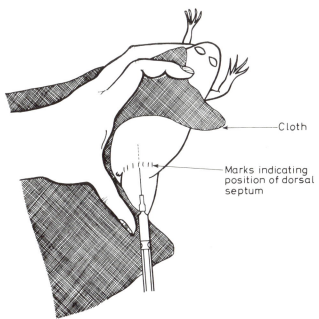

Cloth

Marks indicating position of dorsal septum

Fig. D10. *Method of handling* Xenopus *for injection*

*Materials required for second injection*
> 1 ampoule containing 1 cm³ distilled water
> 1 ampoule containing 500 units Pregnyl
> One 2 cm³ disposable syringe

*Technique for second injection*—Dissolve the Pregnyl in the distilled water as described for the primer injection. Draw up all the solution (1 cm³) into the syringe and inject 0·6 cm³ into the female and 0·2 cm³ into the male. Discard the remaining 0·2 cm³.

Leave the two toads together in one of the small tanks three-quarters full of water.

It is important that the two toads are not disturbed until spawning is completed.

When all the spawn has been laid the toads should be moved back into the stock tank while the spawn may be left in the small tank.

*Care of the tadpoles during development*—The water in the tank containing the spawn should be kept at 23°C. Fill a second small tank with water and maintain it at 23°C.

Two to three days after spawning the tadpoles should begin to hatch and they may then be transferred to the second tank. Use a wide mouthed teat pipette to transfer the tadpoles.

*Feeding*—The tadpoles are herbivorous and may be fed on a suspension of nettle powder. It is important to avoid giving too much food as an excess will start to decompose and foul the water. A little should be given daily or every other day, the amount being increased as the tadpoles grow.

When metamorphosis is taking place the tadpoles will cease to feed on the nettle powder and start to need a carnivorous diet. At this stage the toads will be too small to be fed on chopped earthworms but they may be fed on *Daphnia* or *Tubifex* or on very finely minced pieces of raw beef liver.

*Cleaning*—Once a week the tadpoles should be transferred to a tank of clean water which has been standing for at least 24 h. When the tadpoles have grown too large to be transferred in a teat pipette they should be carried in a small aquarium net. The empty tank should be well cleaned, filled with water and left to stand.

## 3.10. *Mus musculus*

The inheritance of eye colour and coat colour of mice may be used to demonstrate the Law of Segregation and the Law of Independent Assortment. During these experiments some features of the life cycle should be observed.

Experiments may be devised to show some aspects of the behaviour of mice (perhaps by using a maze) and to investigate their physiology, for example measurements may be made of oxygen consumption and carbon dioxide production.

*Source*—In addition to the two suppliers listed at the beginning of the section another source is:—

> Department of Education and Science Laboratories,
> Ivy Farm,
> Knockholt,
> Sevenoaks,
> Kent.

These suppliers keep stocks of mice whose genotypes are known.

If you are going to use your mice for genetic experiments it is inadvisable to buy them from a pet shop as you may introduce unwanted genes.

*Distinguishing characteristics of the male and female*

| Mature male (buck) | Mature female (doe) |
| --- | --- |
| (1) Genital papilla is 1·25 cm from the anus<br>(2) Outline of testes visible<br>(3) No teats visible | (1) Genital papilla is 0·625 cm from the anus<br>(2) No testes present<br>(3) Paired teats of mammary glands visible on ventral surface of abdomen |

*Container*—Two types of polypropylene breeding cage are illustrated (*Plates D5* and *D6*). They are obtainable from:—

> Philip Harris Ltd.,
> 63, Ludgate Hill,
> Birmingham, 3.

Plate D5. *Breeding cage for mice*

The Cambridge cage is also obtainable from:—
Cope and Cope Ltd.,
57, Vastern Road,
Reading,
Berkshire.

Both types of cage are light to handle, easy to clean and they may be autoclaved. Food and water cannot be contaminated with litter or nesting material. Water is supplied from a bottle provided with either a bakelite teat or capillary tubing. Food is placed on a special part of the wire mesh lid.

Care should be taken to see that litter or nesting material is not pushed up the teat or tubing allowing water to escape from the bottle and wet the box.

In the Cambridge mouse cage, designed by Dr. M. E. Wallace, there is a metal sheet which is placed over part of the wire mesh lid and which serves to provide a cover for the nesting area.

*Litter and nesting material*—Sawdust or peat moss are both suitable as litter materials. 1·25 cm layer of either should be spread over the whole floor of the box. Dry hay, wood shavings or clean shredded paper will make good nesting materials and a handful of one of these substances should be placed at the end of the box away from the teat of the water bottle.

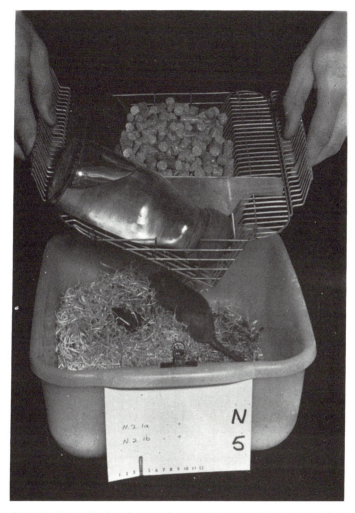

Plate D6. *Cambridge breeding cage for mice (Courtesy of Dr. M. E. Wallace)*

*Feeding and watering*—The most convenient method of feeding is to give a cubed diet. The diets stocked and recommended by the biological suppliers are complete balanced diets so that no supplementary food need be given. With experience you will see how much food to give each mouse every week (*see Part L7*).

Check the water bottles each day to make sure that there is sufficient water available and that the teats are not blocked with nesting material.

*Cleaning*—With this type of cage the soiled litter material may be scooped out about once a week and fresh material may be substituted without disturbing the nesting area.

During this process care should be taken to see that mice cannot jump and fall from the cage possibly injuring themselves and escaping. It is convenient to carry out the cleaning operation by standing the cage in a large cardboard box.

Providing that there are no newly born mice in the nest, about once a month the box should be cleaned thoroughly. All the contents should be removed and the box washed with hot water containing a mild disinfectant. The cage should then be rinsed and dried.

*Handling*—Mice should be lifted by the tail held about half way along by the thumb and forefinger. The front feet should be supported (*see Fig. D11*). Young mice should be lifted bodily from the

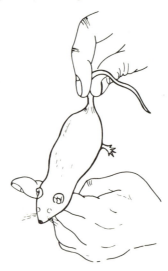

Fig. D11. *To show correct way of lifting a mouse*

nesting area if they have to be handled but care must be taken to ensure that they do not become cold when removed from the nest. It is inadvisable to disturb a new litter as this may cause a doe to abandon her young.

*Parasites and diseases*—Certain diseases are communicable to man thus strict hygiene should be observed when handling mice.

Mice may be parasitised by mites, fleas and lice. The best treatment is to dust the mice with Gammexane powder and disinfect the cages. DDT is poisonous to mice and should not be used.

*Tumours*—These may be visible as swellings. The mouse should be killed.

*Paratyphoid*—The symptoms are excessive diarrhoea. This is an infectious disease and any mouse with the disease or one which has been in contact with the disease should be killed, the bedding burnt and the cage disinfected.

*Fighting*—Occasionally caged adult mice will fight with the result that they may become injured.

We have found that the best method of treatment is to isolate the injured mouse and allow any surface wounds to heal without the application of an antiseptic. If a mouse is badly damaged and in discomfort as a result of a fight it is probably best to kill the mouse.

# Genetics

## 1. TERMS USED IN THIS SECTION

**Allelomorph (allele)**  One of a series of genes each of which may be found at the same site (locus) on a particular chromosome.

**Autosome**  A typical chromosome which does not determine the sex of the organism.

**Back-cross**  A cross between a hybrid and a member of the parent stock.

**Dihybrid**  An organism resulting from a cross between two parents who differ in two distinct characters.

**Dominant**  A feature possessed by one parent which shows itself in the first generation masking an alternative feature derived from the other parent.

**$F_1$**  First filial generation. The offspring which result from a first cross.

**$F_2$**  Second filial generation. This generation is derived from a cross between 2 members of the $F_1$ or in some cases self fertilisation of one organism from the $F_1$.

**Gene**  A part of a chromosome responsible for the development of a feature.

**Genotype**  The genetic constitution of an organism as represented by symbols.

**Heterozygote**  An organism which bears alternative forms of a gene. A hybrid. The dominant feature alone may show so that there is no outward difference between this form and an organism carrying two similar genes for the dominant feature, i.e. a homozygote.

**Homozygote**  An organism which bears identical genes for a given feature. A homozygote is described as a pure breeding organism, all the gametes produced will contain identical genes for the feature under consideration.

| | |
|---|---|
| **Independent assortment** | This Mendelian concept states that during the formation of gametes, genes are distributed independently of one another so that two features which were present together in one of the parents need not be associated in the offspring. This principle has important exceptions, *see* **linkage.** |
| **Linkage** | Genes carried on one particular chromosome are said to be linked. They will tend to remain associated through several generations unless the linkage group is severed by the breakage and consequent crossing over of chromosomes during meiosis. |
| **Mendelian ratio** | The proportions of each phenotype resulting from a cross between two heterozygous organisms when one or two pairs of genes are considered. |
| **Monohybrid** | An organism which is heterozygous for a single pair of genes. |
| **Mutant** | An organism carrying a gene different from any in the parent stock. |
| **Phenotype** | The observable features of an organism. Similar phenotypes may or may not have a similar genetic constitution, e.g. two black mice may appear identical but their genotypes may be different. |
| **Recessive** | A feature of one parent which is masked in the hybrid offspring by the alternative dominant feature provided by the other parent. |
| **Segregation** | The separation of each member of a pair of genes during the formation of gametes. |
| **Sex chromosome** | A chromosome whose presence or absence determines the sex of an organism. The sex chromosomes are usually referred to by letters. The two sex chromosomes in the female *Drosophila* are $XX$ and those in the male are $XY$. |
| **Sex linkage** | A characteristic controlled by a gene which is carried on a sex chromosome is said to be a sex linked character. |
| **Wild type** | The typical form of an organism as found in nature. The genetic symbol for a wild type is $+$. |

## 2. THE VALUE OF CHI-SQUARED $\chi^2$

In experiments where it is possible to make a prediction about the relative numbers of different classes (in these cases, phenotypes) expected in a progeny, it is useful to be able to find out how far the actual numbers obtained may deviate from the expected numbers before the prediction must be assumed to be a faulty one.

$\chi^2$ is a measurement of deviation and the value may be calculated using this equation:

$$\chi^2 = \Sigma \frac{(o - e)^2}{e}$$

where $\Sigma$ = summed for all classes

$o$ = observed number (the number of a particular phenotype obtained in a generation)

$e$ = expected number (the expected number of the phenotype. This may be found by summing the numbers of each phenotype produced and dividing the total by the number relevant to your predicted ratio)

### INTERPRETATION OF THE MEASUREMENT OF DEVIATION

Values of $\chi^2$ have been tabulated for different degrees of freedom.

The number of degrees of freedom ($n$) is the number of classes (phenotypes) minus one.

Once $\chi^2$ has been calculated and the number of degrees of freedom are known the probability of obtaining a deviation of the order calculated can be found by consulting the table of $\chi^2$.

Probability ($p$) in the table is shown as a fraction but is usually expressed as a percentage. If the probability is read as higher than 0·05 (5%) this means that the deviation between observed and expected numbers cannot be regarded as significant and there is no reason to doubt the original predictions. However if the probability is read as less than 0·05 (5%) there is good reason to suspect the validity of the prediction.

### THE APPLICATION OF $\chi^2$

Example: 4 distinct phenotypes were observed in an $F_2$ obtained from a cross between two dihybrids of *Drosophila*. The expected ratio of phenotypes was 9:3:3:1.

| Phenotype | Observed number(o) | Expected number (e) | $o - e$ | $(o - e)^2$ | $(o - e)^2/e$ |
|---|---|---|---|---|---|
| Long wing, grey body | 68 | 68·04 | −0·04 | 0·0016 | 0·000024 |
| Vestigial wing, grey body | 19 | 22·68 | −3·68 | 13·54 | 0·5970 |
| Long wing, ebony body | 28 | 22·68 | 5·32 | 28·30 | 1·2480 |
| Vestigial wing, ebony body | 6 | 7·56 | −1·56 | 2·433 | 0·3218 |
| Total | 121 | 120·96 | 0·00 | | 2·166824 |

$$\chi^2 = 2·166824$$

For 3 degrees of freedom (4 phenotypes − 1) and the above value for $\chi^2$, probability read from the table of $\chi^2$ lies between 0·8 − 0·5 or between 50% and 80%. Therefore, the observed results fit closely with the expected results.

## 3. BREEDING EXPERIMENTS WITH *DROSOPHILA*

(For methods of culture, identification of sex, and anaesthetising procedure, *see Part D3.7*).

Main features of the life cycle of *Drosophila*
(Cultures maintained at 25°C)

| Day | Stage of life cycle | Notes |
|---|---|---|
| 1 | Mating | A cross set up on Friday should produce a new generation of adults in the middle of a week |
| 2 | Eggs laid | |
| 3 | Hatching from eggs | |
| 4 → 6 | Larval period | |
| 7 | Puparia formed | |
| 7 → 10 | Pupal period | Release parents from culture tube |
| 10 → onwards | Adults begin to emerge | Peak of emergence approximately on Day 12 |

## CHARACTERISTICS OF SOME STRAINS OF *DROSOPHILA*

The wild type ( + ) of *Drosophila* is distinguished by grey body colour, wings which extend beyond the posterior tip of the abdomen and red eye colour. Each of these features is subject to mutation. Cultures of *Drosophila* showing one or more mutant features are obtainable from the biological suppliers listed in *Part D3* and *D3.7*.

| *Mutants* | | *Characteristics* |
|---|---|---|
| Vestigial wing | (*vg*) | Short wings with a slightly crumpled appearance |
| Miniature wing | (*m*) | Wings shorter than in the wild type but longer than vestigial wings |
| Curled wing | (*cu*) | Wings curved upwards at 25°C |
| White eye | (*w*) | White eyes |
| Bar eye | (*B*) | Eye has the form of a narrow vertical bar |
| Scarlet eye | (*st*) | Scarlet eye. (Compare with the red of the wild type) |
| Ebony body | (*e*) | Black body. (Not to be confused with black body which is controlled by a different gene) |
| Yellow body | (*y*) | Yellow body |
| Ebony body vestigial wing | (*e–vg*) | Two mutant features present |
| White eye miniature wing | (*wm*) | Two mutant features present |

When ordering stocks of *Drosophila* from the suppliers keep to the conventional method of describing the various strains using the same names and symbols as used by the supplier.

We suggest that you use a microscope to examine the appearance of a member of any new mutant culture received. Make diagrams, coloured if necessary, to show relevant features. These diagrams may be used as references later if there is confusion over phenotypes.

### POINTS FOR CONSIDERATION WHEN MAKING A CROSS

(1) If virgin females are needed for a cross they must be segregated from the males during the 8 h after hatching. Distinguishing the sexes is not easy until the flies are 1 h old.

(2) It is a good plan to have a slight excess of males in the culture tube when making a cross, this ensures that all females will be fertilised at roughly the same time.

(3) Four females in a culture tube should give sufficient progeny to show significant results.

(4) Remember to allow time for the whole generation to emerge before completing the scoring. Some mutants, e.g. vestigial wing, have a slightly longer life cycle than the wild type at 25°C.

(5) In some vestigial wing stock the larvae have a higher mortality than wild type larvae.

## MONOHYBRID INHERITANCE

3.1. A CROSS BETWEEN A WILD TYPE AND A MUTANT WITH EBONY BODY

*Materials required*—Culture of wild type
Culture of ebony flies
Anaesthetising equipment
Fine soft haired paint brush
2 freshly prepared culture tubes for immediate use
2 culture tubes prepared about 12 days after the initial cross has been set up
Binocular microscope or mounted lens

*Procedure*

*DAY 1.   Early morning*—Remove all adults from the 2 culture tubes, either kill them or add to the stock culture.

   *Later in the day* (between 2 and 8 h later)—Obtain newly emerging adults from both culture tubes.

*Set up the 2 culture tubes*
(1) Place 6 male wild type and 4 virgin female ebony flies in one culture tube. Cover the tube and keep at 25°C.

(2) Place 6 male ebony and 4 virgin female wild type in the other culture tube. Cover the tube and keep at 25°C.

   Label both tubes to show the contents and date of making the cross. For example the label for culture tube (1) should be:—

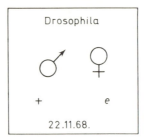

*DAY 8*.    Examine both culture tubes. Larvae and puparia should be visible if the tubes have been kept at 25°C. Remove and kill the parent flies.

*DAY 10–DAY 20*.    Anaesthetise all adult flies which emerge.

Examine and record the body colour and sex of every fly.

Keep the records separate for each culture tube.

The flies which emerge are the $F_1$ generation.

Keep about 12 virgin females and 12 males of the $F_1$ generation. Segregate the sexes and place all the virgin females in one culture tube and all the males in another tube.

After examination and recording, the rest of the $F_1$ may be killed.

## FURTHER INVESTIGATION USING THE $F_1$ GENERATION

*Materials required*—Culture tube containing virgin female $F_1$
Culture tube containing male $F_1$
Culture of ebony flies
Culture of wild type flies
3 freshly prepared culture tubes
Anaesthetising equipment
Fine soft haired paint brush
Binocular miscroscope or mounted lens

3.2. A CROSS BETWEEN THE $F_1$

*DAY 1*.    4 virgin females and 4 males of the $F_1$ are placed together in a culture tube. The tube should be covered, labelled and kept at 25°C. Dispose of parent flies when pupae are visible.

*DAY 10–20*.    Anaesthetise all flies to emerge. As before record the body colour and sex of each fly. This is the $F_2$ generation.

(If you wish to proceed further to investigate the genotype of some of the members of the $F_2$ remember to segregate enough virgin females and males in separate culture tubes before disposing of the $F_2$.)

### 3.3. A BACK CROSS BETWEEN THE $F_1$ AND EBONY BODIED FLIES

*DAY 1.* 4 virgin females of the $F_1$ and 4 males with ebony body colour (from the same stock as the parent of the $F_1$) are placed together in a culture tube. The tube should be covered, labelled and kept at 25°C.

Dispose of parent flies when puparia are visible.
*DAY 10–20.* Anaesthetise, examine and record body colour and sex of each fly to emerge.

After recording dispose of flies.

### 3.4. A CROSS BETWEEN THE $F_1$ AND THE WILD TYPE

*DAY 1.* 4 males of the $F_1$ and 4 virgin females of the wild type (from the same stock as one of the parents of the $F_1$) are placed together in a culture tube. The tube should be covered, labelled and kept at 25°C.

Dispose of parent flies when puparia are visible.
*DAY 10–20.* Anaesthetise, examine and record body colour and sex of each fly to emerge.

After recording dispose of flies.

*Answer the following questions*

(1) Are the phenotypes of the $F_1$ similar in each culture tube? What do you think was the purpose of setting up both tubes?
(2) Is there any numerical relationship between the phenotypes obtained in the $F_2$?
(3) What proportions were there of each body colour in the offspring of the back cross between the $F_1$ and the ebony flies?
(4) How would you determine by experiment the genotypes of selected offspring from the cross between the $F_1$ and the wild type flies?
(5) Using symbols to represent the genes controlling body colour and using the information obtained from your investigations, give the genotypes of the following:—

The ebony body parent and the wild type parent
The $F_1$
The $F_2$—all the possible genotypes
The offspring of the back cross between the $F_1$ and the ebony flies
The offspring of the cross between the $F_1$ and the wild type flies.

*Further work*

By using cultures of other mutants, e.g. flies with vestigial wings or scarlet eyes, in place of the ebony bodied flies, perform investigations 3.1–3.4, and determine whether these mutant characteristics are inherited in a similar way as the characteristic of ebony body.

## INDEPENDENT ASSORTMENT

3.5. A CROSS BETWEEN FLIES WITH VESTIGIAL WINGS (AND GREY BODY COLOUR) AND FLIES WITH EBONY BODIES (AND WINGS OF NORMAL LENGTH)

*Materials required*—Culture of *Drosophila* with vestigial wings and grey body colour
Culture of *Drosophila* with ebony bodies and wings of normal length
Anaesthetising equipment
Fine soft haired paint brush
2 freshly prepared culture tubes
Binocular microscope or mounted lens

*Procedure*

*DAY 1.   Early Morning*—Dispose of all adults from the 2 culture tubes.
   *Later in the day*—Obtain 6 male vestigial wing and 4 virgin female vestigial wing flies, 6 male ebony and 4 virgin female ebony flies.
*Set up the 2 culture tubes*
(1)  Place 6 male vestigial wing and 4 virgin female ebony flies in one culture tube. Cover and label the tube and keep at 25°C.
(2)  Place 6 male ebony and 4 virgin female vestigial wing in the other tube. Cover and label the tube and keep at 25°C.
*DAY 8.*   If puparia are visible, remove and dispose of the parent flies.
*DAY 10–20.*   Anaesthetise all adult flies which emerge.

Examine and record the wing length (either vestigial or long), body colour and sex of each fly.

Keep separate records for each culture tube.

Isolate in 2 separate culture tubes 12 virgin females and 12 males from this $F_1$ generation.

When all the $F_1$ have been examined kill those flies no longer needed.

FURTHER INVESTIGATION USING THE $F_1$ GENERATION

*Materials required*—Culture tube containing virgin female $F_1$
Culture tube containing male $F_1$
Culture of vestigial wing flies
Culture of ebony body flies
3 freshly prepared culture tubes
Anaesthetising equipment
Fine soft haired paint brush
Binocular microscope or mounted lens

*Procedure*

Follow the same procedure for setting up the crosses as described previously. Examine and record the phenotypes on the days suggested.

*Set up the following crosses*

3.6. A cross between the $F_1$ to give the $F_2$.
3.7. A back cross between the $F_1$ and the ebony bodied flies.
3.8. A back cross between the $F_1$ and the vestigial winged flies.

*Answer the following questions*

(1) If each phenotype is defined only by the two features of wing length and body colour how many distinct phenotypes are present in the $F_2$?
(2) Are the phenotypes present in a definite ratio to one another? If so what is this ratio?
(3) If from your results vestigial wings appear to be inherited independently of ebony body colour, what does this lead you to

suggest about the distribution in the nucleus of the genes controlling these features?

(4) Give all the possible genotypes of the $F_2$.


## INDEPENDENT ASSORTMENT AND LINKAGE

3.9. A CROSS BETWEEN BLACK BODIED, VESTIGIAL WINGED FLIES AND EBONY BODIED FLIES HAVING WINGS OF NORMAL LENGTH

*Materials required*—Culture of black bodied vestigial winged flies
(obtainable from Philip Harris Ltd.)
Culture of ebony bodied flies
1 freshly prepared culture tube
Anaesthetising equipment
Fine soft haired paint brush
Binocular microscope or mounted lens


*Procedure*

*DAY 1.   Early morning*—Remove and dispose of all adults from the 2 culture tubes.

*Later in the day*—Obtain 6 male black bodied, vestigial winged flies and 4 virgin female ebony bodied flies and put them together in a freshly prepared culture tube. Cover and label the tube and keep at 25°C.

*DAY 8.*   If puparia are visible, remove and dispose of the parent flies.

*DAY 10–20.*   Anaesthetise, examine and record the phenotypes of all adults to emerge. Be particularly careful over your description of body colour.

Isolate 6 virgin female and 6 male flies of this $F_1$ generation. Dispose of the remaining members of the $F_1$.


3.10. A CROSS BETWEEN THE $F_1$ TO GIVE THE $F_2$ GENERATION

*Materials required*—6 virgin female $F_1$
6 male $F_1$
1 freshly prepared culture tube
Anaesthetising equipment
Fine soft haired paint brush

*Procedure*

Place the 6 virgin female $F_1$ and the 6 male $F_1$ in a culture tube. Cover, label and keep at 25°C.
*DAY 8.*    If puparia are visible remove the parent flies.
*DAY 10–20.*    Anaesthetise, examine and record the phenotypes of all the adults to emerge. As in 3.9 be particularly careful with your observation of body colour.

Unless you wish to proceed further with investigations of the genotypes of $F_2$ dispose of these flies once you have made your records.

*Answer the following questions*

(1) Suggest a genotype for the $F_1$. (Remember that 3 features are involved.)
(2) How many distinct phenotypes are present in the $F_2$? In what ratio to one another do they occur?
(3) Is the ratio of phenotypes in the $F_2$ the one you would have expected from a theoretical calculation?
(4) Which characteristics are linked together (controlled by genes carried on the same chromosome)? Which characteristics are inherited independently of one another?
(5) From data obtained as a result of investigations 3.1 → 3.10, give examples of dominant and recessive characteristics and the genotypes of a homozygote and a heterozygote.

## SEX LINKAGE

The sex chromosomes in the female *Drosophila* are similar in shape and are given the symbols *XX*. The sex chromosomes in the male are different and are given the symbols *XY*, the *Y* chromosome has a slightly hooked appearance.
    The wild type of fruitfly has red eyes. A mutant form has white eyes.
    If we assume that the mutant form of the gene for white eyes (*w*) is carried on that part of the *X* chromosome which has no homologous region in the *Y* chromosome and that this one gene can overshadow the effect of the other genes for eye colour, then a male who carries the mutant gene on his *X* chromosome will always be obvious by the possession of white eyes, his genotype may be represented thus *Xw Yo*.

A female fruitfly will only have white eyes if she is homozygous for this gene, i.e. $Xw\ Xw$.

Test to see if these assumptions are correct by performing the following investigations.

*Materials required*—Culture of white eyed flies
Culture of red eyed flies (+ culture)
2 freshly prepared culture tubes for immediate use
2 culture tubes prepared for use about 12 days after the first crosses have been made
Anaesthetising equipment
Fine soft haired paint brush

### 3.11. CROSS BETWEEN RED EYED FEMALES (+) AND WHITE EYED MALES

*DAY 1.   Early morning*—Dispose of all adult flies from the two cultures.

*Later in the day*—Obtain at least 3 virgin red eyed females and place them in a culture tube with 6 white eyed males. Cover and label the tube and keep at 25°C.

*DAY 8.*   If puparia are visible dispose of the parent flies.

*DAY 10–20.*   Anaesthetise, examine and record the phenotype and sex of each fly to emerge.

*Keep at least 3 virgin female flies and 6 male flies from the $F_1$.*

### 3.12. CROSS BETWEEN THE $F_1$

*Procedure*

In one of the culture tubes place together 3 virgin female and 6 male $F_1$ flies. Keep the culture tube at 25°C. After the appropriate length of time (see previous investigations) and after having removed the parent flies record the phenotypes and sex of all the newly emerging flies.

*Answer the following questions*

(1) What proportion of the $F_2$ are red eyed flies and what proportion are white eyed?

(2) What proportion of the red eyed flies are females and what proportion males?

(3) Which sex are the white eyed flies?

(4) Do these results support the initial hypothesis?


3.13. CROSS BETWEEN RED EYED ( + ) MALES AND WHITE EYED FEMALES

*Procedure*

*DAY 1.* Follow the same procedure as before but this time set up the cross between 3 virgin white eyed females and 6 red eyed ( + ) males.

Leave them together in a culture tube at 25°C.

*DAY 10–20.* As before record the sex and phenotype of each fly to emerge.

*Keep at least 3 virgin female flies and 6 male flies from the $F_1$.*


3.14. CROSS BETWEEN THE $F_1$

*Procedure*

In one of the culture tubes place together 3 virgin female and 6 male $F_1$ flies. Keep the culture tube at 25°C. After the appropriate length of time and after having removed the parent flies record the phenotypes and sex of all the newly emerging flies.


*Answer the following questions*

(1) What proportion of this generation of flies is red eyed and what proportion is white eyed?

(2) Are there both male and female white eyed flies? Is the same true for red eyed flies?

(3) Do your results support the initial assumption that the $X$ chromosome carries the gene for white eyes?


## 4. BREEDING EXPERIMENTS WITH MICE

(For notes on the care of mice in a school laboratory and notes on the features which distinguish the males from females *see Part D3.10.*)

NOTES ON THE LIFE CYCLE OF MICE (AVERAGE TIMES GIVEN)

| Day | Notes |
|---|---|
| 1 | Mating of parents |
| 1–(19)21 | Gestation period (some variation) |
| 21 | Birth of litter. 4–9 is average number in a litter |
| 27 | 6 days after birth the teats of the young females are visible. Sexing possible |
| 35 | Weaning of young begins |
| 41–43 | Weaning complete and young may be removed from parents |
| 62–64 (9 weeks) | Young are sexually mature at 6 weeks old. Males and females should be segregated |
| 64 onwards | Breeding possible. In optimum conditions a new litter may be produced once every 4 weeks for at least 8 months |

## SOME OF THE STRAINS OF MICE AVAILABLE FROM SUPPLIERS

(Philip Harris Ltd keep all the strains described here. Other suppliers from whom some of the strains may be obtained are given in Part D.)

Strains of mice are kept which are homozygous for all gene pairs and which have one or more pairs of genes present in the mutant form.

When ordering mice it is essential to use the same symbols for genotypes as used by the supplier and not to use other descriptions of coat colour, eye colour, ear length, etc.

## KEY TO THE SYMBOLS USED BY SUPPLIERS

Where a stock is described by one pair of symbols only it may be assumed unless otherwise stated that the remaining genes are present as homozygous dominants. Check with the supplier if you are in any doubt.

### Examples

Agouti mice listed as $AA$ have the genetic formula $AABBCCDDPP$.
Black mice listed as $aa$ have the genetic formula $aaBBCCDDPP$.

THE MAIN AUTOSOMAL GENES WHICH DETERMINE COAT AND EYE COLOUR

| Dominant allele and effect | | Other alleles in the series and their effects | |
|---|---|---|---|
| *C* | Colour.<br>Presence enables the formation of pigment. | *cc* | No colour, regardless of other genes present. Albino, i.e. with white coat and pink eyes. |
| | | *c^e c^e* | Extreme chinchilla. Pale coffee coat and black eyes. Yellow band of agouti absent. |
| | | *cc^e* | This genotype produces a mouse with a nearly white coat and black eyes. |
| *A* | Agouti.<br>Called after the agouti, a large rodent, whose coat the wild mouse coat resembles. Hairs of the wild mouse have black pigment granules along their length except near the tip where there is a band of yellow pigment. The *A* allele allows the formation of the yellow band; the colour of the non-yellow part of the hair is determined by genes outside this allelic series. | *A^y* | Yellow coat. *A^y A^y* is lethal before birth. *A^y* is dominant to *A*, *a^t* and *a*. |
| | | *a^t a^t* | The *a^t* allele produces a non-agouti dorsal coat with yellowish belly hairs, i.e. 'black and tan' in the absence of other colour mutants *a^t* is dominant to *a*. |
| | | *Aa^t* | This genotype has the agouti dorsal coat of the wild mouse but a paler or more yellow belly. |
| | | *aa* | Non-agouti coat. No yellow band but non-yellow pigment extends along length of hair. The coat is black in the absence of other mutants. |
| *B* | Black.<br>When *C* is present and *A* is absent, i.e. *BBCCaa* genotype a black coat is produced. | *bb* | Black pigment is changed to brown. Yellow pigment is not affected. In the presence of *C* and *a*, a 'chocolate' brown mouse is produced. In the presence of *A*, a 'cinnamon' is produced. Brown base to hairs with yellow tips. Speckled appearance. |
| *D* | Dense pigmentation. | *dd* | Dilution. Causes a reduction in the number of pigment granules produced and an irregular distribution of those present. Yellow pigment only slightly affected.<br>*ddaa* is a slatey-blue.<br>*ddAA*, slatey-blue with a yellowish bloom due to *AA*. |
| *P* | This allele allows full coat colour and pigmentation of eyes. | *pp* | Pink eyed dilution.<br>Yellow pigment not affected.<br>Black pigment changed to a pale grey-blue. *ppAA* has almost yellow hair due to *A*, *ppaa* has grey-blue coat known as lilac.<br>Absence of pigment in eye gives pink eye. |

'Blue' mice listed as *dd(aa)* have the genetic formula *aaBBCCddPP*.
'Chocolate' mice listed as *bb(aa)* have the genetic formula *aabbCC-'DDPP*.
Albino mice listed as *cc(aa)* have the genetic formula *aaBBccDDPP*.
Pink eyed dilution mice listed as *pp* have the genetic formula *AABBCCDDpp*.
Extreme chinchilla mice listed as $c^e c^e(aa)$ have the genetic formula *aaBBc^e c^e DDPP*.

It is advisable when ordering a particular strain to specify the agouti part of the genotype, i.e. whether you require *AA* or *aa*.

The symbol + is used to denote a non-mutant or normal feature, $++$ denotes a homozygote, $+$ denotes a heterozygote.

E.g. a cross between a homozygous black and a homozygous brown may be symbolised: $++(aa) \times bb(aa)$
and from this the heterozygote produced would be: $+b(aa)$.

In these examples + means *B*.

### REFERENCE SKINS

It is a good plan to build up a set of preserved and mounted skins labelled to show the genotype of the mouse from which the skin was obtained. The coat colour of each genotype is then available for comparison with the coat colours of mice produced in later breeding experiments.

### THE PREPARATION OF REFERENCE SKINS

The following method for preserving and mounting skins on card is based on a method suggested by the British Museum.

*Materials required*—White card—*Size of card for each skin*: The length of the mouse from nose to end of tail. 50 mm width or approximately half the length of the head and body, tapered at one end to fit the anterior end of the mouse (*see Plate E1*).
Magnesium carbonate
Borax
Paradichlorobenzene
(Wire and cotton wool or splints of garden cane) optional

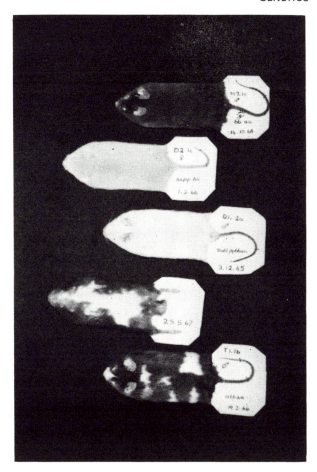

Plate E1. *Mounted skins of mice for reference*

Fuse wire or thread and needle
Small scissors, scalpel and forceps
Rubber solution

*Skinning*—Use magnesium carbonate as an absorbent powder to clear up blood shed during skinning.

Cut through the skin at the back of one knee and continue the incision dorsal to the anus and ventral to the tail, round to the other knee.

Loosen the skin round the incision by cutting through connective tissue.

Cut rectum and urinogenital tract.

Expose base of tail; grip vertebrae with forceps, grip skin at base of tail and pull to extract vertebrae. This stage is difficult and leaving the tail attached to the skin although aesthetically desirable is not necessary.

Cut through muscle and bone of each knee.

Peel skin forwards, turning it inside out and cut fore legs as hind legs.

Cut ears close to skull, cut neatly inside eyelids and sever skin from head by cutting inside lips.

*Preparing the skin*—Remove any flesh, fat or glandular tissue from the skin and clear flesh from leg bones.

Rub borax over the inner surface of the whole skin.

*Mounting*—If the skin of the tail has been left intact insert either a wire wound round with cotton wool or a bamboo splint.

Hold the inverted skin and card nose to nose and roll the skin on to the card. The card should fit the skin without the skin being wrinkled or stretched.

Secure the tail to the dorsal surface of the card with fuse wire or thread. Attach the hind feet to the ventral side of the card.

The fore legs may be stuck to the skin with rubber solution.

Brush the fir with a stiff brush.

*Labelling*—List the following information on the dorsal surface of the card:—

Reference number of mouse

Sex of mouse

Genotype

Date of birth

On the ventral surface of the card record the date of preparation of the skin.

*Storage*—The mounted skins should be kept in boxes with crystals of paradichlorobenzene to deter insect pests.

KEEPING RECORDS

It is recommended that all stocks are checked once a week so that the information recorded on loose leaf paper and on the cage cards is kept up to date.

## A LOOSE LEAF SYSTEM FOR PERMANENT RECORDS

This description of a simple record system for breeding experiments with mice is based on the method described by Dr. M. E. Wallace (Department of Genetics, University of Cambridge) in *Sch. Sci. Rev.* (June, 1965) **160**, 646 and on personal communication.

*Requirements*—Loose leaf file
　　　　　　　　Squared paper ($\frac{1}{4}$ inch)
　　　　　　　　Coloured card dividers to separate the records for
　　　　　　　　　different projects
All obtainable from:—Mr. G. A. de Vall,
　　　　　　　　　　Waterlow & Sons Ltd.,
　　　　　　　　　　85–86 London Wall,
　　　　　　　　　　London, E.C.2.

*Procedure*

Each mating is recorded on a separate sheet of paper.

All the pages carrying data connected with one project are given the same reference letter, e.g. *M* indicating that in this case the project deals with monohybrid inheritance. This letter is placed at the top left of the page.

Each page is numbered in order with the number placed by the side of the letter, e.g. the second mating in this project would have *M2* at the top left side. The pages are routinely kept in serial order, but can be removed temporarily and rearranged for collation of data. Terminated mating records are removed to a separate loose leaf file.

The following information on each mating should be recorded.

At the top right hand side the date on which the male and female were placed together.

It is only necessary to record a few facts about the male and female and you should be consistent (for easy reference) over the order in which these are placed. The reference number, genotype and date of birth of each mouse should be recorded on a single line: details for the female first, then the male, near the top centre of the page.


*The top part of a page in the records might look like this:—*

*M2*                                                               **18.2.69**
            *M1.2b* ♀ *Dd aa* **21.11.68**
            *M1.1d* ♂ *Dd aa* **16.10.68**


*Key* (Unnecessary in records)

*M2*        = mating number.
**18.2.69** = date on which male and female were placed together.
*M1.2b*     = the reference number of the female.
            *M*1 = the mating number. Her parents were the first to
                   be mated in this project.
            *2* means that she was a member of the second litter.
            *b* = her specific symbol in that litter.
*Dd aa*     = the genotype of both male and female.
**21.11.68** = date of birth of female.
*M1.1d*     = the reference number of the male, i.e. the mating number
              followed by the information that he was born in the first
              litter and that *d* is his specific symbol.
**16.10.68** = date of birth of the male.


   Records are kept of each litter to be born to this pair. The first details may be written on one line and should show the litter number (first, second, third, etc.) the number of mice in the litter and the date of birth.
   When it is possible to sex the young mice each mouse should be given a letter.
   When their phenotypes are clear, these should be noted by the side of the appropriate sex symbols. If there are several mice of the same sex and phenotype, some distinguishing mark may become necessary at some time. These should be noted at the time they are made against the relevant letter in the records, using appropriate symbolism (e.g. *L* = left ear clipped, *T* = tail clipped, *LT2* = second toe of left hind foot clipped, *G* = dorsal, green felt pen mark).
   The genotype, derived from the phenotype of each young mouse and the genotype of its parents is extended as further data are obtained from later breeding experiments.

*The sample page might look like this after two litters had been born to the parents*

*M2*                                                                18.2.69

*M1.2b* ♀ *Dd aa* **21.11.68**
*M1.1d* ♂ *Dd aa* **16.10.68**

**1  5  31.3.69**
*a* ♀ *D a*
*b* ♀ *d a*
*c* ♀ *D a*
*d* ♂ *D a*
*e* ♂ *d a*

**2  6  4.5.69**
*a* ♀ *D a*
*b* ♀ *D a*
*c* ♀ *D a*
*d* ♂ *D a*
*e* ♂ *d a*
*f* ♂ *d a*

If this method of recording data is followed there will be room on the page, if symbols are used, to indicate the fate of each mouse. For example the complete records for the second litter on page *M2* might look like this:—

**2  6  4.5.69**
*k*.20.5.69  *a* ♀ *D aa*
          *b* ♀ *D aa*   *L M3*
          *c* ♀ *D aa*    *M4*
*k*.20.5.69  *d* ♂ *D aa*
          *e* ♂ *d aa*   *M3,4*
          *f* ♂ *d aa*  *S*

*Key*

| | |
|---|---|
| *k* | = killed and the date. |
| *L* | = left ear clip, made to distinguish identical females when placed together in the trio. |
| *S* | = stored. |
| *M3* and *M4* | = the mating numbers of a mated trio. |

Further matings are needed in this project if more information is required about the genotypes of some of the progeny. The next sheet in the records might start in the following way:—

*M3*                                                                        **18.6.69**

        *M2.2b ♀ D aa    L 4.5.69*
        *M2.2e ♂ dd aa      4.5.69*


CAGE CARDS


Cards 11·875 × 9·375 cm (4¾ × 3¾ in) quadrille ruled on one side only, with numbers 1–12 on the bottom squares are obtainable from:—

        Mr. K. Brett,
        Galloway & Porter,
        30, Sidney Street,
        Cambridge.

To avoid confusion over the contents of cages used in breeding experiments, the front of each cage should carry a card clearly labelled to give current information about the mated mice inside the cage.

Each breeding female should be given a card and this should always be attached to the cage which she occupies. If a trio of mice is used the cage will carry two cards; on some cages these may be placed one behind the other, the mating number of the card behind projecting to the right of the front card so that it is clearly seen.

Each card is set out as shown (information about litters other than that given below should not be written on the card: this and other non-current information is too cumbersome).

The mating number is written large on the right hand side; the reference numbers and phenotype or minimum genotype, of the female and her mate are half way down and to the left; and, starting from the bottom left hand corner and extending along the base of the card, are the numbers 1–12.


*The use of coloured plastic paper clips*

These may be attached to the card and show, on a quick inspection, the condition of the female e.g. whether she is pregnant, has a new litter or an old one etc.

The position of the clips is altered as necessary after the routine examination of the cages.

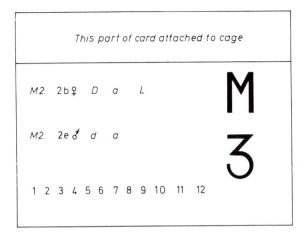

Two different pieces of information can be conveyed by one coloured clip, as it may be attached to the card with either the broad or the narrow side visible.

Four different coloured clips, red, blue, green and yellow are amply sufficient for use with these cards.

*Specimen card with three clips in position*

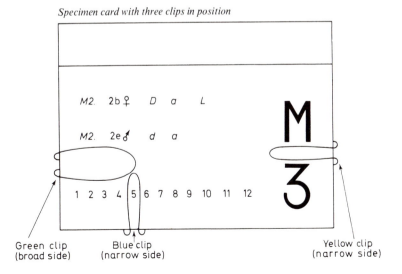

SUGGESTED POSITIONS AND MEANINGS OF CLIPS

| Colour | Visible side | Position | Meanings |
|--------|--------------|----------|----------|
| Red | Broad | Right hand side, between symbol and number. | Female with male. Litter expected within the week. |
| Red | Narrow | Right hand side, between symbol and number. | Female with male. ♀ probably pregnant. |
| Yellow | Broad | Right hand side, between symbol and number. | Female separated from male. ♀ pregnant. |
| Yellow | Narrow | Right hand side, between symbol and number. | Female separated from male. ♀ nursing. |
| Blue | Broad | Base of card, over relevant number. | Number covered indicates litter size. Broad side indicates age of litter (e.g. old enough to classify, or old enough to separate from mother). |
| Blue | Narrow | Base of card, over relevant number. | The litter is younger than the age specified above. |
| Green | Broad | Left hand side, centre. | Cage should be inspected daily. (Perhaps the female is a poor mother and litter will need fostering.) |
| Green | Narrow | Left hand side, centre. | Disease suspected in litter or adults: decision on culling required. |

## MONOHYBRID INHERITANCE

To obtain sufficient progeny from each type of cross in one school term it is suggested that 2 trios of mice are used simultaneously.

A trio consists of 2 females and 1 male. Each trio occupies 1 cage but 1 female may be removed to another cage just before her litter is due if the original cage is too small to accommodate 3 adults and 2 litters.

SOME SUGGESTED EXPERIMENTS

### Dominance

4.1. Cross between a black and a 'chocolate' mouse.

$$aa \times aabb$$
$$\downarrow$$
$$F_1$$

4.2. Cross between $F_1$ to produce $F_2$.

4.3. Back cross between $F_1$ and 'chocolate' parent.

4.4. Cross between an agouti and a black mouse.

$$AA \times aa$$
$$\downarrow$$
$$F_1$$

4.5. Cross between $F_1$ to produce $F_2$.

4.6. Back cross between $F_1$ and black parent.

*Interpretation*—You should now have examples of dominant and recessive features.

Do the numbers of each phenotype obtained in the $F_2$ fit those expected from a theoretical calculation? Carry out the chi-squared test if necessary.

## Incomplete dominance

4.7 Cross between an albino and an extreme chinchilla.

$$cc \times c^e c^e$$
$$\downarrow$$
$$F_1$$

4.8. Cross between $F_1$ to produce $F_2$.

4.9. Cross between an agouti and a black with tan belly.

$$AA \times a^t a^t$$
$$\downarrow$$
$$F_1$$

4.10. Cross between $F_1$ to produce $F_2$.

*Note the following*—The phenotypes of the $F_1$ in both the above experiments.

The ratio of the phenotypes in the $F_2$.

Does the ratio obtained fit the one which you expected?

Give the genotypes of all phenotypes in the $F_2$.

## DIHYBRID INHERITANCE—Independent assortment

SOME SUGGESTED EXPERIMENTS

4.11. Cross between a 'chocolate' and a blue (Maltese dilution).

$$aabb \times aadd$$
$$\downarrow$$
$$F_1$$

4.12. Cross between the $F_1$ to give the $F_2$.

(At least 3 trios necessary in the $F_1$ for a reasonable chance that all phenotypes will be present in $F_2$.)

4.13. Cross between a short eared mouse and a pink eye dilution.

$$sese \times pp$$
$$\downarrow$$
$$F_1$$

4.14. Cross between the $F_1$ to give the $F_2$.

(Use at least 3 trios of $F_1$.)

It is necessary to wait until the progeny are 18 days old before classifying for ear size.

*Interpretation*—Interpret the data obtained with the aid of diagrams giving the genotype of each phenotype in the $F_1$ and $F_2$.

Do your results suggest that *b* and *d* are carried on the same chromosome or on different chromosomes?

Does the same apply to *se* and *p*?

## Sex linkage

A gene which is carried on the non-homologous region of the *X*-chromosome is useful to demonstrate linkage between a characteristic and the sex of the mouse. One gene is known as the Tabby gene, *Ta*.

*Source*—Stocks of males and females which carry *Ta* are kept by Philip Harris Ltd.

You should note the difference between the phenotypes of *TaX* (tabby female) and *TaY* (greasy coated male).

SUGGESTED EXPERIMENTS

4.15. Cross between a tabby female and a normal male.

$$TaX \times XY$$

(At least 2 trios should be used.)

4.16. From the $F_1$ mate a normal female with a greasy coated male.

4.17. Cross between a tabby female and a greasy coated male.

$$TaX \times TaY$$

Account for the results you obtain.

# 5.  BREEDING EXPERIMENTS WITH *TRIBOLIUM*

(For method of culture, identification of sex and method of setting up a cross, *see Part D3.8.*)

## SOME OF THE STRAINS AVAILABLE

*Tribolium confusum*

| Phenotype | Genotype |
|---|---|
| Wild type | $+$ |
| Ebony body | $e^2$ |
| Black (semidominant)<br>$\begin{cases}\text{Homozygote, black } bb \\ \text{Heterozygote, bronze } +b\end{cases}$ | $b$ |
| Pearl eye (with darker border) | $p$ |

*Tribolium castaneum*

| Phenotype | Genotype |
|---|---|
| Wild type | $+$ |
| Black body (similar to that of *T. confusum*) | $b$ |
| Jet body | $j$ |
| Sooty body | $s$ |
| Paddle (some fusion of segments of antennae) | $pd(X)$ |
| Pearl eye with darker border | $p$ |
| Microcephalic (reduced cranial region) | $mc$ |

## MONOHYBRID INHERITANCE

SOME SUGGESTED EXPERIMENTS

5.1.  Cross between wild and sooty *Tribolium castaneum*

$$\downarrow$$
$$F_1$$

5.2  Cross between members of the $F_1$ to give $F_2$.

5.3  Cross between wild and black bodied *Tribolium confusum*
↓
$F_1$

5.4.  Cross between $F_1$ to give $F_2$.

Note the phenotypes of the $F_1$ in each experiment and compare with those of the parents.

How many distinct phenotypes are there in each $F_2$ generation?

Are the ratios you obtained those which you expected?

Carry out the chi-squared test if necessary.


## DIHYBRID INHERITANCE

5.5.  Cross between pearl eyed (red-brown body) and sooty body (black eyed) *Tribolium castaneum*.  ↓
$F_1$

5.6.  Cross between $F_1$ to give $F_2$.

From an analysis of the data obtained from this experiment can you say with reasonable certainty that these genes for eye colour and body colour are carried on different chromosomes?

# PART F
# Nutrition

The exercises appropriate to this topic are divided into three sections.

Section I of Part F describes and investigates the ways in which different animals collect their food.

The activity of some digestive enzymes has been examined already in Part B.

Methods for the analysis of some products of digestion are described in Section II of Part F.


## SECTION I   FEEDING MECHANISMS

Feeding mechanisms are related to the diet of the animal, the structure of the parts available for food capture and the habitat of the animal.

One of the main purposes of movement for an animal is to obtain food. An organelle or organ used for locomotion may also be adapted for food capture. The pseudopodia of rhizopods, the cilia of ciliated protozoans and some crustacean limbs all serve a double function of locomotion and food capture.

Animals may take in their food in the form of relatively small particles, comparatively large pieces, or in the liquid state, thus three distinct animal groups are recognisable.


### Microphagous animals

These are animals whose food source is in the form of comparatively small particles. Included as particulate food are bacteria, phyto- or zooplankton or fragments of organic material.

It is characteristic of a microphagous animal to have:
(1) A method of creating and directing a food carrying water current.
(2) A surface to act as a filter for the collection of suspended particles.
(3) In many cases a secretion of mucus to help to bind the particles into a more manageable mass before they are directed into the digestive cavity.

There may be further refinements in the form of sorting mechanisms to eliminate particles of the wrong size.

Many microphagous animals use cilia for creating the feeding current. Examples of ciliary feeders are found in all phyla of the Animal Kingdom except in the Arthropoda and Nematoda.

There are filter feeders among arthropods but in place of cilia, limbs fringed with setae are used by some crustaceans to create a water current and form a straining mesh.

*Chirocephalus*, the fairy shrimp, a member of the Branchiopoda may be taken as an example of a microphagous crustacean. This animal swims on its back being propelled forwards by the beating of the serially arranged thoracic limbs. As the animal moves forward, water is sucked into the inter-limb spaces and any suspended food material is left behind on the setae of the thoracic limbs. Food particles are combed off by very fine setae, they are then directed into the median food groove by water currents, entangled in mucus and pushed into the mouth by appendages round the mouth.

Filter feeding is not restricted to the invertebrate phyla and examples are found throughout the vertebrate classes. The herring traps zooplankton with its gill rakers, the tadpoles of many amphibians filter food from the water passing through the gill pouches. Flamingoes use efficient filters for selecting the required type of food from the water in which they are wading. Whalebone whales use a filter of keratin plates, hanging from the roof of the mouth, to strain off euphausiids, shrimp like crustaceans.

## Macrophagous animals

These are animals whose food source is in a relatively large mass. The mass of food either may be ingested whole or is broken up into smaller pieces before it is taken into the alimentary canal. The external breakdown of food may be accomplished either mechanically or chemically. The mechanical breakdown of food is carried out with the use of appendages adapted for cutting, tearing or crushing the food before it is taken into the mouth.

The chemical breakdown of food occurs as a result of the secretion of digestive enzymes on to the food.

*Calliphora* larvae partly digest their food externally before ingestion and some spiders inject a salivary protease into their prey to liquefy the victim before it is pumped by a muscular pharynx into the alimentary canal. This latter type of animal might equally well be classified as a fluid feeder.

## Fluid feeders

These are animals whose food source is in a fluid state. The origin of the liquid food may be either plant e.g. cell sap or nectar or animal e.g. blood, coelomic fluid or intestinal contents.

The first part of the alimentary canal of these animals is adapted for suction and in some the mouth parts may be specialised for piercing the outer covering of plants or animals in order to reach the enclosed tissue fluids.

Those animals which feed by sucking blood may secrete an anticoagulant to ensure a free flow of blood.

Some fluid feeders e.g. the cestodes lie immersed in their food which is absorbed over the whole body.

Some insect orders are particularly well adapted for a fluid diet. The Homoptera and Heteroptera have lance like mouth parts used for piercing and sucking. Aphids, members of the Homoptera use their mandibular and maxillary stylets for piercing plant tissues, cell sap being drawn up through a food canal formed by the shaped inner edges of the two maxillae. *Cimex*, the bed-bug, a member of the Heteroptera has mouth parts adapted for piercing human skin and withdrawing blood. The Lepidoptera and certain of the Diptera have mouth parts adapted for suction without any modifications for piercing.

There are fluid feeders in other phyla. The nematodes, the opisthobranch group of molluscs and the Hirudinea are groups whose members utilise a liquid diet.

INTRODUCTION TO PRACTICAL WORK

Feeding mechanisms may be studied in all the animals kept in a school laboratory. Find out which food substances are selected by the animal in its natural habitat and if possible provide this type of food in the laboratory. (*See Part D* for the food requirements of some laboratory animals.)

Before you start to observe feeding in any animal it is essential that you are familiar with the anatomy of the animal, particularly those parts involved in food capture. Knowledge acquired as a result of dissection, sectioning and/or staining followed by microscopical examination if necessary will be very valuable.

Where it is difficult to see the very small food particles utilised by microphagous animals, use a dilute solution of a stain to colour the particles.

In the case of microscopic animals where good lighting is essential for clear observation, be very careful that heat from the light source is not directed on to the animal. Avoid microscopes with built in illumination.

Food may be rejected by macrophagous animals if they have recently been fed, time will be saved if your observations of feeding mechanisms take place on the presentation of food to relatively hungry animals.

## 1. THE INGESTION OF PARTICLES AND FORMATION OF FOOD VACUOLES IN *PARAMECIUM*

*Materials required*—*Paramecium* culture
                      Cavity slides
                      Cover slips
                      Microscope
                      Pipettes
                      Powdered borax carmine
                      Yeast cells
                      Congo red
                      10% solution of methyl cellulose

*Procedure*

To slow down the movements of *Paramecium* add a drop of 10% methyl cellulose to a drop of the culture on the cavity slide.
(1) Add a very small amount of powdered borax carmine to the drop of culture and methyl cellulose on the slide. Cover with a cover slip and examine under H.P. of a microscope.

Are the borax carmine particles ingested? If they are ingested does this occur when *Paramecium* is moving or when the animal is resting? If the particles are ingested watch to see if food vacuoles are formed.

(2)  Preparation of coloured yeast cells—3 g yeast
                                    30 mg congo red
                                    10 cm³ distilled water
Boil gently together for 10 min and cool before adding to the culture.
    Congo red is an indicator for the range pH 3·0 to pH 5·0:
        at pH 3·0 congo red is blue-violet
        at pH 5·0 congo red is red-orange
Add a drop of the coloured yeast suspension to a drop of the
*Paramecium* culture on a cavity slide. Cover with a cover slip and
examine under a microscope. You should observe the ingestion of
yeast cells and formation of food vacuoles. Have you any indication
that there is a change in the pH of the vacuolar contents during the
path of the vacuole through the cytoplasm?

## 2. OBSERVATION OF FEEDING AND THE PASSAGE OF FOOD PARTICLES IN *DAPHNIA*

*Daphnia* is a microphagous crustacean. The feeding apparatus
consists of 5 pairs of trunk limbs, whose combined function is the
collection of small particles of food, and the mouth parts which
transfer the food to the mouth.

Each of the trunk limbs is differentiated for a particular function
but all 5 pairs work together to form an efficient food gathering
mechanism. The co-ordinated beating of all the limbs is responsible
for drawing in water under the carapace so that it passes over the
limbs and through the inter-limb spaces.

The first and second limbs have long setae which may help in
preventing particles which are too large from entering the food
groove. The third and fourth limbs are each provided with a fringe
of many setae which act as the filters on which the food particles are
trapped. The fifth pair of limbs is without the fringe of setae and
does not play any part in the filtering mechanism, the main function
of this pair of limbs is to help in the maintenance of the feeding
currents.

*Materials required*—Culture of *Daphnia* (*Simocephalus* is also suit-
                     able)
                     Dried yeast
                     Neutral red
                     Cavity slides
                     Cover slips

Microscope
Pipette
Cotton wool

*Preparation of coloured yeast cells*—Add some water to a small amount of dried yeast to make a suspension of yeast cells.

Prepare a 0·02% aqueous solution of neutral red and use this to colour the yeast cells.

Neutral red is an indicator for the range pH 6·8 to pH 8·0:

pH 6·8 red
pH 7·2 light red (rose)
pH 8·0 orange/yellow

*Procedure*

Place *Daphnia* in a drop of culture solution in a cavity slide. Add a few cotton wool fibres to slow down the movement of *Daphnia*. Add a small drop of the coloured yeast suspension and examine under a microscope. A lateral view of *Daphnia* will show you the most detail.

Observe the paths of the yeast cells from the liquid to *Daphnia*.

Make an outline diagram of *Daphnia* to show where the yeast cells are drawn in under the carapace.

Note the rhythmically beating thoracic limbs within the carapace. Can you distinguish any of the setae on these limbs?

Is there any sign that yeast cells are being concentrated in the region of the filter prior to being taken into the gut?

Focus carefully on the anterior region near the point at which the gut contents are first visible. Watch for several minutes to see if you are able to observe yeast cells being taken into the gut. After some minutes you should be able to see an accumulation of the coloured cells in the first part of the gut. Is there any colour change in the cells during their passage through the gut?

## 3. FEEDING IN *HYDRA,* INVESTIGATION OF THE STRUCTURE AND FUNCTION OF THE TENTACLES

*Materials required*—*Hydra* culture
Permanent stained preparations of *Hydra*, whole mounts and sections
*Artemia* eggs
Sodium chloride

Calcium chloride
Sodium bicarbonate
Methyl green acetic
Cavity slides
Cover slips
Pipettes
Microscope
Fine mesh net
1 shallow container for the culture of *Hydra*
1 shallow container for the *Artemia* eggs

## THE CULTURE OF *HYDRA*

*Hydra* must be kept either in water from the source from which they were collected or in a specially prepared culture medium. Tap water should not be used either alone or as the solvent of the culture medium.

Stock solutions suitable for the culture of the brown hydra, *Hydra oligactis* (syn. *Hydra fusca* and *Pelmatohydra oligactis*) are prepared as follows:—

*Stock solution 1*: Sodium chloride   133·0 g
                                Calcium chloride   26·6 g
                                Distilled or de-ionised water to 1000 cm³
*Stock solution 2*: Sodium bicarbonate 38·0 g
                                Distilled or de-ionised water to 1000 cm³

Use 10 cm³ of each solution and dilute up to 4 litres with distilled water.
Use this culture medium to fill shallow dishes to a depth of about 3 cm.

## THE USE OF *ARTEMIA NAUPLII* AS FOOD FOR HYDRA

*Artemia* (brine shrimp) eggs may be obtained from:—

Charlton's Pet Shop,
15, Yorkshire Street,
Morecambe,
Lancashire.

Prepare a salt solution by dissolving 3·5 g NaCl in 1 litre of water. Pour some of this solution into a shallow container to give a depth of 1–2 cm.

Scatter the eggs on to the surface of the solution.

Do not move the container after this.

Leave for at least 48 h.

*Artemia* nauplii swim towards light. If part of the container is darkened the larvae will collect in the part open to the light and may then be collected from this area using a fine mesh net. While they are in the net wash off the salt solution using the hydra culture medium. To maintain a stock of *Hydra*, feed twice a week with freshly hatched nauplii. After feeding decant the culture medium and replace with fresh solution.

### THE STRUCTURE OF THE TENTACLES

Examine stained preparations of *Hydra* which show the cellular structure of the tentacles. Identify and note the distribution of the nematocysts. Compare the structure of the tentacle wall with the general body wall.

Examine a living hydra placed in a little culture medium in a cavity slide. Cover with a cover slip and observe under high power. Carefully examine the structure of the tentacles and the hypostome.

Use a 5% aqueous solution of NaCl or a solution of methyl green acetic to irrigate *Hydra*. Note the discharge of the nematocysts.

### THE CAPTURE AND INGESTION OF FOOD

*Hydra* is a carnivorous coelenterate feeding upon relatively large animals. It selects as prey certain living crustaceans e.g. *Daphnia*. Either *Daphnia* or *Artemia* nauplii are readily ingested by a starved hydra in the laboratory.

Use either a binocular microscope or a microprojector to enable you to see the sequence of events after the presentation of food to a hydra in a small amount of culture medium.

Are you able to describe how the prey is caught? After the capture of the prey watch the response of all the tentacles. How many tentacles are involved in directing the prey to the mouth? Are you able to see if the mouth opens on the capture of the prey?

## 4. THE FEEDING MECHANISM OF LOCUSTS

The locust belongs to the macrophagous group of animals. It is a herbivore which is able to utilise many different types of vegetation.

The success of this insect depends to a large extent on the efficient functioning of the feeding apparatus.

Some idea may be obtained of the devastation which can be caused by a swarm of locusts if a single locust is observed feeding; a considerable amount of plant material is consumed in a very short time.

To appreciate the significance of each pair of mouth parts you will need to examine them in detail to enable you to relate structure to function.

EXAMINATION OF THE MOUTH PARTS OF *LOCUSTA*

*Materials required*—1 preserved locust per student
1 living locust for feeding observations
Hand lens or dissecting microscope
Fine scissors, forceps, mounted needles
Watch glasses
Microscope slides
5% potassium hydroxide
Test tube
Bunsen burner

*Additional requirements if a permanent preparation of the mouth parts is to be made*—Industrial methylated spirits (70%, 90% and 'absolute alcohol')
Xylol
Canada balsam
Square cover slip
Pieces of glass cut in strips the same length as the cover slip

*Procedure*

Examine the head of the locust using *Fig. F1* to help you identify the position of the mouth parts.

You should notice that the labrum extends downwards covering the anterior faces of the mandibles. You will see the black edge of the left mandible to the left side of the labrum. Identify the visible parts of the maxillae and labium.

Use a mounted needle to lift up the labrum to enable you to see the powerful mandibles. Separate the maxillae and labium with a needle and examine with the help of a lens.

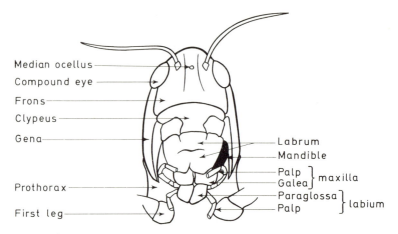

Fig. F1. *The front of the head of locust as seen from below to show the position of the mouthparts*

If you wish to study the structure of the individual mouth parts you will have to remove them from the head of the locust.

*Procedure for the removal of the mouth parts*

(1) Cut through the neck membrane to separate the head from the thorax.
(2) Place the head in 5% KOH in a test tube and boil for 30 min.
(3) Rinse in water.
(4) With fine forceps and fine scalpel gently remove in order the **labrum,** the **labium,** the **maxillae,** the **mandibles** and the **hypopharynx.** Make a clean cut through each muscle attached to the mouth parts.
(5) Lay out the isolated mouth parts in water and make drawings to show the structure of each type.

*To make a permanent preparation of the mouth parts*

Treat all the mouth parts in the same way.
(1) Dehydrate in 70% alcohol 5 min
(2) Transfer to 90% alcohol 5 min
(3) Transfer to 'absolute' alcohol 5 min
(4) Transfer to fresh 'absolute' alcohol 5 min
(5) Clear in xylol 10–15 min
(6) Mount in Canada Balsam.

The cover slip will be unstable unless it is supported round the edge. Glass strips, the same thickness as the mandibles, may be placed under the edges of the square cover slip to keep it horizontal.

*Answer the following questions*

(1) How many clearly defined regions are there on each mandible? Suggest a function for each region you have described. Watch the mandibles and labrum in action when a locust feeds.

(2) On each maxilla you will see a pair of claws. What do you think their function might be?

(3) Observe a locust feeding and carefully watch to see the action of the maxillary and labial palps. The palps are jointed, flexible and relatively long when compared with the other sections of the maxillae and labium. Do these structural features give you any clue as to their particular function?

Devise ways of investigating your ideas.

(4) When a locust feeds are the limbs used either to hold or to tear vegetation?

*Suggestions for further work*

Design your own experiments to try to find out if these two statements are true.

(*a*) The locust is an indiscriminate herbivore.

(*b*) One locust will consume its own weight or more than its own weight of food daily.

## 5. SUGGESTIONS FOR THE OBSERVATION OF FEEDING IN *XENOPUS*

When watching an animal feed unless one is particularly observant and analytical some of the finer details of food capture and ingestion may be missed.

To help you to make critical observations it may be as well before you begin work to prepare a list of questions which when answered will provide you with useful information about the feeding mechanism.

If *Xenopus* is kept in your school laboratory you will know that the adults are carnivorous.

A hungry toad feeds rapidly so that you will need to watch very carefully when food is presented.

Food should be given with the minimum of disturbance as once *Xenopus* are disturbed they take time to settle.

*Materials required*—One adult *Xenopus* (due to be fed)
 Chopped earthworm/chopped liver
 Pestle and mortar
 Teat pipette
 Thread
 Dummy earthworm/piece of twig or something similar

DETECTION OF THE PRESENCE OF FOOD

Name the sense organs which may be available for the perception of the presence of food. It is possible to distinguish between the importance of the various receptors which you list? How would you find out if touch receptors play any part in the process of locating food?

Tie a dummy earthworm or piece of plant material to the end of a thread and dangle it in front of a hungry *Xenopus*. What response is given?

Crush an earthworm or piece of liver with a little water in a mortar. Draw up some of the liquid extract and gently pipette it into the water some way in front of *Xenopus*. Is any response made to the presence of the extract? If a response is made suggest which receptors are involved.

THE CAPTURE AND INGESTION OF FOOD

How is food captured? Do the fore or hind limbs play any part in handling the food and transferring it to the mouth? If they do, do you consider that they are well adapted for this function?

Does the food appear to be held within the buccal cavity or are there signs that it is swallowed immediately?

Do the eye balls appear to play any part in the swallowing process?

If possible examine the mouth and the buccal cavity of a preserved *Xenopus* and contrast with *Rana*.

COMPARISON OF THE LARVAL AND ADULT *XENOPUS*

Adult *Xenopus* are macrophagous and carnivorous but the tadpoles may be classified as microphagous herbivores, as they feed on phytoplankton in their natural surroundings. They are true filter feeders drawing in a current of water through their mouths and sieving out fine suspended plant cells before the water is expelled.

*Materials required*—*Xenopus* tadpoles
                                    Nettle powder
                                    Binocular microscope
                                    Watch glass
Use a binocular microscope to examine a 1 week old tadpole in water containing a little nettle powder.

Make a list of the structural and behavioural features associated with the two types of feeding mechanism.

In this way you will be made aware of the extent of some of the changes which take place during the metamorphosis of *Xenopus*.

## 6. FLUID FEEDERS

A number of animals in this group are endoparasites whose habits are not easy to observe.

However there are a number of fluid feeding insects which are readily obtainable particularly in the warmer months of the year.

### Examples of some commonly occurring fluid feeders

*Musca domestica*—the housefly
*Calliphora* sp.—the blowfly
*Culex pipiens*—a mosquito
*Apis mellifica*—the honey bee
*Aphis* sp.—the aphid

NOTES FOR THE EXAMINATION OF THE MOUTH PARTS

The mouth parts can only be examined in detail after they have been displayed as a result of dissection. This requires considerable skill in the case of the relatively small insects listed here.

It is possible to buy from the main biological suppliers permanent preparations of the mouth parts of a number of fluid feeding insects.

You should examine as many of these as possible.

*Note the following points*

In an insect which obtains its food by suction it is essential that some of the mouth parts form a tube or tubes up which the fluid may be drawn. With the aid of textbook diagrams, try to distinguish the mouth parts which form the food canal(s).

Which of the insects have mouth parts which are specialised for piercing plant or animal tissues? This type of mouth part will be needle like.

Are there accessory structures, for example, in the form of jointed palps? If there are, have they the same form in all the mouth parts you examine?

*Observation of the feeding insect*

When watching an insect feeding, the glass container housing the insect should be supported at eye level. This will enable you to obtain a side view of the feeding insect. Use a lens to see the head clearly.

If you watch an aphid or a housefly feeding you will see that the position of the mouth parts relative to the rest of the head is altered from the resting position. You will see a change in the silhouette of the head.

When either insect is resting, the proboscis is not obvious from a lateral view, when the insect feeds the proboscis is extended towards the food.

Describe what you can see of the proboscis and its functioning.

# SECTION II  ANALYSIS

## 1. CHROMATOGRAPHY AS A TECHNIQUE FOR ANALYSIS

During digestion proteins are hydrolysed to produce amino-acids. The amino-acids are then absorbed through the wall of the gut and assimilated. Proteins can be hydrolysed in the laboratory by the action of mineral acids. A rapid and reasonably simple method of identifying the products of such a hydrolysis is to use chromatography.

## Principle of chromatography

The principle of chromatography was discovered by the Russian botanist Tswett (1872–1919). He extracted the chlorophyll pigments from leaves with petrol and passed the solution through powdered chalk carefully packed into a long vertical glass tube. As the solution percolated through the column it was noticed that the component pigments were moving through the chalk at different speeds. When pure petrol was added to the top of the column a complete separation occurred. The component pigments appeared at intervals on the column of chalk.

## Adsorption chromatography

The form of chromatography used by Tswett in the experiment described above, is termed adsorption chromatography. The substances in the solution are adsorbed on to the surfaces of the substance making up the column through which it is passed. The separation denotes that certain substances have differing intensities of adsorption; those that are most strongly adsorbed appearing first on the column.

## Partition chromatography

If the medium used in the column holds some water and the solvent containing the substances to be separated is immiscible or only just miscible with water, separation occurs by a method other than adsorption. Provided that the substances contained in the solvent are equally soluble in water, when the solvent is added to the top of the column, some of each of them will pass to the water being held by the medium packing the column. The distribution of these substances between the solvent and medium will be such that

$$\frac{\text{concentration of a substance in the solvent}}{\text{concentration of the substance in the water}} = \text{a constant}$$

if the temperature and conditions of the experiment remain the same. This method of separation is termed partition chromatography and the constant for each substance is termed the partition coefficient of the substance.

Instead of a column, paper may be used to hold the stationary water phase. (Filter paper holds about 15% water and is often used, although separation does not occur entirely by partition.)

## Thin layer chromatography

Partition and adsorption chromatography may also be carried out on thin layers of adsorbent material which have been applied to a thin glass plate or plastic sheet as a support. The medium may hold water and here separation occurs by partition or it may be dry and then separation occurs by adsorption.

In partition chromatography (here liquid/liquid), the water held by the medium is termed the stationary phase, the solvent moving down the medium (descending chromatography) or moving up the medium (ascending chromatography) is termed the moving phase. The solute is separated into differing components which move at varying rates and remain in the water at certain positions according to their differing partition coefficients. The relative rate of movement of the solute to solvent is termed the $Rf$ factor (for Rate of flow). For given substances and the same chromatographic system the $Rf$ is constant for each substance. The $Rf$ factor is expressed as

$$\frac{\text{Distance moved by solute}}{\text{Distance moved by solvent}} \; (\textit{see Fig. F6})$$

## Paper chromatography

A general procedure is given below.

*Apparatus needed*—A developing tank, for example a large gas jar (30–40 cm high), Kilner jar, or battery jar (suggested size 10 cm diameter by 20 cm deep, obtainable from Griffin and George). The tank should be large enough to take sufficient solvent to run the chosen length of paper and have a closely fitting lid, bung or cork.

If a strip of paper is to be used, a device should be arranged to hold the strip of paper and prevent it from touching the side of the tank and if possible to allow the paper to be lowered into the solvent when the top has been fitted prior to the commencement of the development of the chromatogram (*see Fig. F2*).

If a number of spots are to be applied for comparison purposes and the paper chosen is too wide for the tank, then the paper may be formed into a cylinder and the two edges stapled together (*see Fig. F3*). To achieve a fine separation where a large number of components are known to be present, the chromatogram may be run twice. The second run is at right angles to the first and a different solvent is used (*see Fig. F4*). This is termed two dimensional chromatography.

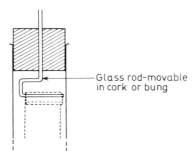

Glass rod–movable
in cork or bung

Fig. F2. *Filter paper holder for one dimensional development*

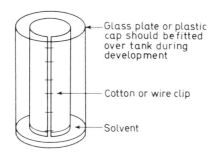

Glass plate or plastic
cap should be fitted
over tank during
development

Cotton or wire clip

Solvent

Fig. F3. *Arrangement of paper and tank for two dimensional chromatography*

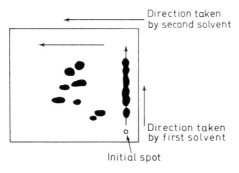

Direction taken
by second solvent

Direction taken
by first solvent

Initial spot

Fig. F4. *Paper after two dimensional development*

*Paper*—Whatman filter papers Nos. 1, 3 and 3MM are suitable grades for chromatography. No. 4 has a fast flow rate. For any specific separation, different grades should be tried. Rolls of paper of varying widths may be purchased (Griffin & George).

*Application of spot*—The solution of substances which are to be separated should be carefully spotted on to the paper. The spot should be confined to an area not exceeding 4 mm in diameter. This may be achieved by using a simple micro-pipette which can be made by drawing out a piece of glass tubing (*see Fig. F5*). The solution is

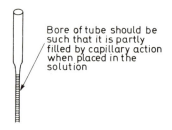

Bore of tube should be such that it is partly filled by capillary action when placed in the solution

Fig. F5. *Micropipette*

drawn up into the fine tube of the pipette by capillary action. The micro-pipette should be held carefully particularly if the solvent is very volatile as the heat of the hand may cause it to spurt from the end of the pipette. A small amount of the solution should be applied to the end of the paper (a pencilled circle may be drawn on the paper to aid accuracy in positioning the spot) and allowed to dry (*see Fig. F6*). Further amounts may be added to increase the concentra-

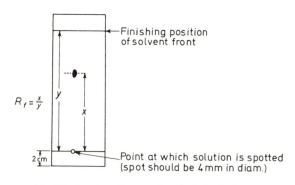

$R_f = \frac{x}{y}$

Finishing position of solvent front

Point at which solution is spotted (spot should be 4 mm in diam.)

Fig. F6. *Position of spot and method of calculating* Rf

tion, but this may not be necessary if the spot locating reagents are very sensitive.

*Development of the chromatogram*—The chosen solvent can be allowed to descend or ascend the paper. It is perhaps, easier to use the ascending method though it may have drawbacks if separation is slow.

In the ascending method, the solvent is poured into the bottom of the tank. A depth of about half an inch will be most adequate. The tank should be left so that its atmosphere is completely saturated with the solvent vapour. The paper which has been spotted should be carefully lowered into the tank and the bung inserted. Using the rod (*see Fig. F2*) carefully lower the paper into the solvent so that the spot is left about 1 cm above the level of the solvent. The end of the paper should be at least 1 cm below the level of the solvent. The apparatus should then be left undisturbed. The condition of the solvent front should be examined from time to time. The experiment should be stopped and the paper withdrawn from the tank just before the solvent front reaches the top of the paper. The time depends on the grade of paper used and the solvent employed.

When separation has occurred, it may be necessary to locate the spots or streaks on the paper by means of chemical reagents which react with the substances to make them coloured and thus visible. The reagents can either be sprayed on to the paper using a scent spray or prepared aerosol or dipped into a tray containing the reagent.

1.1. A CHROMATOGRAPHIC EXAMINATION OF THE PRODUCTS OF PROTEIN HYDROLYSIS

*Materials required—*
   (*a*)   *One dimensional development*
         30 (or 40) cm gas jar with bung or cork bored to take glass rod (as in *Fig. F2*).
         Strip of No. 1 Whatman filter paper 34 (or 44) cm by 3 cm. The end of the filter paper to be held at the top of the jar should be turned over and stapled (*see Fig. F2*).
   (*b*)   *Two dimensional development*
         Suitable glass tank or battery jar (10 cm in diameter by 20 cm deep).
         Large cork to fit tank or, for jar of dimensions given above, a plastic lid of a coffee tin.

Square of filter paper cut from large filter paper or sheet filter paper, 18 cm × 18 cm.
Cotton with needle or small stapler.
A saturated aqueous phenol solution (this may be prepared by adding 20 cm³ of distilled water to 80 cm³ of white unoxidised phenol in a separating funnel. The solution should be well shaken and left for two hours before use. Just before use in the developing tank add 0·5 cm³ ammonium hydroxide per 100 cm³ of phenol).

*For either of the above methods the following will also be required—*
Small pestle and mortar
Micro-pipette (*see* introduction)
Evaporating basin
50 cm³ conical flask with a loosely fitting cork
250 cm³ measuring cylinder
250 cm³ beaker
Small filter funnel and paper
5N HCl solution (prepared by carefully adding 43 cm³ concentrated HCl to 100 cm³ distilled water)
Solvent (prepared by adding 40 cm³ of n-butanol to 10 cm³ glacial acetic acid and 20 cm³ of distilled water)
Dried egg albumen
10 cm³ of a 50% aqueous isopropanol solution
Ninhydrin aerosol (B.D.H. from Philip Harris). Ninhydrin may be prepared for use in a scent spray as a 0·2% solution in isopropanol.

*Procedure*

Weigh out 25 mg of egg albumen and powder in a mortar. Put the powdered albumen in the conical flask and add very carefully 10 cm³ of the prepared 5N HCl solution, shake carefully making sure that the albumen does not go lumpy. The flask should then be very lightly corked and placed in an oven for 12 h at 100°C. (Alternatively, the solution can be refluxed for 15 h.) At the end of the period remove the solution from the oven. Place the solution in an evaporating basin and evaporate the excess acid. When cool add 50 cm³ of distilled water and stir. Filter the solution then evaporate the filtrate. To the residue add 2 cm³ of the aqueous isopropanol. This solution is then ready for spotting on to the chromatography paper as described in the general procedure.

If technique (*a*) is to be employed the butanol/acetic acid/water solvent should be used. If technique (*b*), the first solvent to be used is the phenol solution followed by the butanol/acetic acid/water for the second run. The butanol/acetic acid/water solvent will take about 12 h to move up a 30 cm paper and the phenol about 10 h to move up the 18 cm paper.

At the completion of the runs the paper should be carefully dried (in an oven at 80°C or with a hair dryer), and then the ninhydrin sprayed on. The paper should then be returned to the oven and dried at 100°C or very carefully (the spray is inflammable) held over a bunsen flame. Ninhydrin fumes should be avoided and if the paper is to be heated over a flame, the operation should be carried out in a fume cupboard. Hold the paper carefully to avoid perspiration from

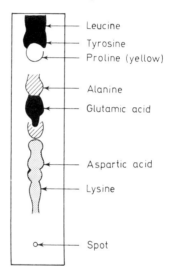

Fig. F7. *Appearance and possible position of separated amino acids*

the fingers getting on to the paper. The wearing of rubber or plastic gloves is recommended or the hands should be well washed beforehand.

The paper should develop in a few minutes and should then be removed from the oven or heat for the identification of the spots which are coloured through blue to lilac (*see Fig. F7*).

If a chromatogram is made with known solutions of amino-acids (made up in a 1:1 solution of ethanol and 0·1N HCl) and each run separately together with a sample spot of the prepared solution for the identification of the amino-acids present and calculation of *Rf* factors, the chromatogram should be run once only. The solvent used is butanol/acetic acid/water.

*Further work*

(*a*) Instead of hydrolysing protein with an acid try the action of pancreatin. A mixture of soluble proteoses, peptones, polypeptides and amino-acids will usually result but a chromatogram would be worthwhile and a control of known amino-acids also could be run for comparison.

(*b*) Sucrose may be hydrolysed with N HCl and the result applied to a filter paper and a chromatogram run. When a developing reagent is used two spots one of glucose and one of fructose will appear. These may be identified by running a control chromatogram.

(*c*) The preparation of chromatoplates in thin layer chromatography is a fairly difficult procedure. Shandon Scientific Co. Ltd. (65 Pound Lane, Willesden, London, N.W.10) produce a kit (Unikit No. 2) which allows complete experiments including the loading of the plate and developing to be attempted. Kodak Ltd. (Kirkby, Liverpool) have marketed both plastic sheets already coated and several models of the Eastman 'Chromagram' developing apparatus (Model 104 is the simplest developing apparatus but cannot be used for two dimensional work).

# Respiration

## 1. THE PRODUCTION OF CARBON DIOXIDE DURING RESPIRATION

*Materials required*—Three conical flasks or three Dreschel bottles
Glass and rubber tubing for connections (*see Fig. G1*)
One bottle with wide mouth
Stoppers for bottle and flasks if used
Filter pump or large aspirator with connections
0·5N sodium hydroxide solution
Lime water
Animal e.g. mouse or earthworm

*Procedure*

Assemble the apparatus as shown in the diagram (*see Fig. G1*). Into flask *A* pour sufficient NaOH solution so that the end of the long tube is covered. Flasks *B* and *C* should both contain lime water. The animal should be placed in the wide necked bottle. Flask *A* is connected to flask *B* and this is in turn connected to the bottle containing the animal. This bottle is connected to flask *C* which is joined to the aspirator or filter pump.

The pump is operated or the aspirator tap is turned on so that a slow stream of air is drawn through the system. The NaOH solution in flask *A* will absorb any $CO_2$ in the incoming air and this will be shown by the lime water in flask *B* remaining clear. During the next twenty minutes or so, observe the condition of the lime water in flask *C*. If $CO_2$ is expired by the animal, the lime water in flask *C* should gradually turn milky.

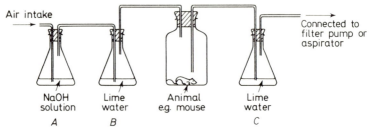

Fig. G1. *Apparatus to demonstrate the production of carbon dioxide as a result of respiration*

*Further work*

If a mouse has been used in the experiment, substitute an earthworm or vice versa. With the air stream being drawn in to the system at the same rate, note the time it takes the lime water to go milky in each case. If there is a difference in time, how can you explain this?

## 2. MEASUREMENT OF RESPIRATORY QUOTIENT

In aerobic conditions, when nutrients are oxidised, the amount of oxygen needed alters according to the chemical structure of the nutrient. As a result of oxidation carbon dioxide and water are released.

By recording the amount of oxygen consumed and carbon dioxide released information can be obtained which will give some indication as to the nutrients involved in respiration.

The ratio of the amount of $CO_2$ produced to the amount of $O_2$ used is called the respiratory quotient ($RQ$).

$$RQ = \frac{\text{Volume of } CO_2 \text{ released}}{\text{Volume of } O_2 \text{ used}}$$

e.g. Glucose has a $RQ$ of 1 from the simplified equation.

$$C_6H_{12}O_6 + 6O_2 = \text{energy} + 6CO_2 + 6H_2O$$

The evaluation of gas exchange can be read either as a volume change using a volumeter or a pressure change as shown by a manometer. When quantitative results are required with either of these methods a correction should be made for changes in atmospheric pressure and temperature. A control apparatus referred to as a thermobarometer should be used to make the necessary corrections.

By the use of two or more volumeters and measuring either oxygen consumption or carbon dioxide release per unit time, it is possible to compare the rate of respiration of organisms (e.g. mouse, frog), or to show how a factor such as temperature will affect the respiration of a living organism.

DESCRIPTION OF VOLUMETER AND ITS USE

The container of the volumeter should be large enough to take the living material and $CO_2$ absorber, if required. The opening should be as wide as possible to allow easy access. Suitable containers are honey jars, large glue jars, Kilner jars, coffee jars and certain specimen jars.

The removable top should have a tight fit and be able to take the side arm tube and pressure equalising tube (refer to diagram). It is recommended that a rubber bung should be used. Bungs can be obtained (Griffin & George) to fill openings up to 110 mm.

When the side arm, pressure equalising tube and/or syringe have been fitted (*see Fig. G2*) and the whole volumeter assembled, the

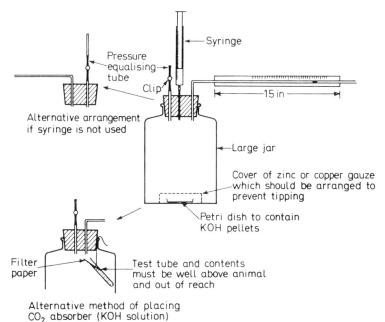

Fig. G2. *Construction of volumeter*

apparatus should be tested to establish that the system is air tight.

By using a syringe the need to find the volume of the side arm will be eliminated, the volume change can be read off by moving the plunger of the syringe to return the index to its starting point.

If a syringe is not used the volume of the tubing may be found (providing the bore is of even diameter), by filling a unit length of the tubing from a syringe and graduating. A pipette can also be used to graduate the tubing.

Providing the bore of the side arm is uniform, for comparative work, recording unit movement of the index will suffice.

The volumeter and control volumeter (thermobarometer) should be assembled as indentical systems.

The index fluid may be prepared from washing up detergent coloured with eosin, coloured soap bubble fluid or by preparing Brodie's fluid (NaCl 4·6 g, sodium tauroglycholate 1 g, distilled water to 100 cm³ adding a crystal of thymol and a little eosin). The index fluid is inserted into the side arm by using a syringe or rubber teat pipette. It can be moved along the side arm by withdrawing the plunger of the syringe acting as pressure equaliser, or opening the clip of the pressure equalising tube (if a syringe is not fitted) and withdrawing some air. A light ruler or a strip of graph paper mounted on card may be placed on the side arm as a suitable scale.

## 2.1. THE MEASUREMENT OF RESPIRATORY QUOTIENT USING A SMALL MAMMAL

*Materials required for each group—*

> 3 volumeter assemblies—one to act as a thermo-barometer
>
> 2 small mammals e.g. mice (these should be the same weight)
>
> An object (e.g. rubber bung) or objects approximately equivalent to the volume of a single mouse
>
> Index fluid
>
> Stop watch
>
> N.KOH solution or KOH pellets
>
> Glass or plastic beads
>
> 3 Petri dishes and zinc gauze covers

*Procedure*

In volumeter 1, place in a Petri dish a quantity of KOH pellets. The gauze cover (which must be stable) is fixed over the dish. A mouse is placed in the volumeter and the bung assembly is inserted without the syringe in place or the clip of the pressure equalising tube being tightened.

In volumeter 2, instead of the KOH pellets an equivalent volume of an inert substance (e.g. glass or plastic beads) is placed in the

Petri dish. The mouse is placed in the volumeter and the bung minus syringe or clip is inserted as before. This volumeter measures overall gas exchange.

In volumeter 3, which will act as the thermobarometer, the procedure as for volumeter 2, is followed except that an inert object equivalent to the volume of a mouse is placed in the container.

In the side arms of all three volumeters the index should be positioned about one third of the way from the open end.

Leave the volumeters a few minutes before inserting the index which should be followed by closing the clips on the pressure equalising tubes or inserting the syringes.

*Interpretation*

Carefully observe the movements of the indices during the next 5–10 min. Irregular movements of the index will be observed if the mouse in a volumeter is fairly active. It is suggested that the mice should previously be made familiar with the confined space of the volumeter. The mice should not be left too long in an air tight volumeter or if they appear distressed.

After the given time (possibly determined by the rapidity of movement of the index of volumeter 1), the syringe plungers should be manipulated to return the indices to their starting places and the volume change, read from the syringe barrel or, the volume change may be read directly from the side arm as a volume, or unit movement.

The *RQ* may be calculated as follows:—

Let the movement of the index for a given time (after applying the correction from the reading of the thermobarometer as an addition or subtraction to both volumeter readings) equal:

in Volumeter 1 $a$ cm³ (or $a$ mm)

in Volumeter 2 $\pm b$ cm³ (or $\pm b$ mm). A movement of the index away from the volumeter indicates more carbon dioxide is released than oxygen used and is a positive reading. A movement of the index towards the volumeter end of the side tube indicates less carbon dioxide is released than oxygen used and is a negative reading.

$$RQ = \frac{a + (\pm b)}{a}$$

*Further work*

If the rate of respiration or *RQ*'s of small invertebrates (e.g. earthworm, woodlouse, spider) are to be studied it is recommended that a small respirometer be used. Such a respirometer is produced by Philip Harris (*see Fig. G3*) and is both simple and convenient to use.

The volume changes may be equated with a pressure change using a manometer. This change may be read directly or the index returned to equilibrium by moving the plunger of the attached syringe and

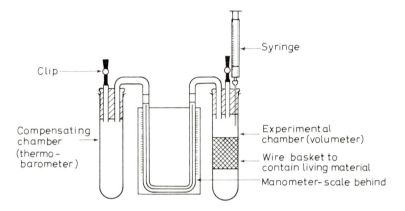

Fig. G3. *Simple respirometer*

the volume read from the scale on the syringe barrel. The empty test tube will act as a thermobarometer and adjustments due to environmental fluctuations will be automatically applied.

The rate for the same animal could be measured at different temperatures by placing the two test tubes in a water bath maintained at a known temperature.

## 3. MEASUREMENT OF THE VITAL CAPACITY OF THE LUNGS OF MAN

The amount of air taken into the lungs and expelled from the lungs during normal quiet breathing movements is termed the **tidal volume.** This volume does not represent the total lung capacity of an individual. By using the inspiratory muscles it is possible to inhale a volume of air to fill the lungs to a near maximum. This amount in excess of the normal tidal volume is termed the **inspiratory reserve volume.** By using the expiratory muscles it is possible to exhale a volume of air in excess of that recently inhaled. This volume represents the **vital capacity** of the individual lungs. An amount of air termed the **residual volume** cannot be removed and remains trapped in the alveoli of the lungs (*see Fig. G4*).

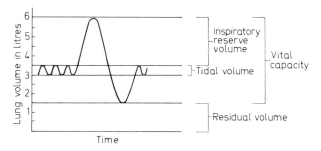

Fig. G4. *Vital capacity in relation to other lung volumes*

*Materials required*—The use of a large sink and adjacent cold tap. It is convenient if a hollow cylindrical drain stopper is used.

6 litre bell jar with tubulure at top. This should be calibrated using paper covered with Sello-tape or with a Chinagraph or Glassrite pencil

A filter pump with attachments (*see Fig. G5*)

Three or four metal blocks (iron stands with the uprights removed are ideal), on which the bell jar will rest in the sink

A mouth piece and attachment

Methylated spirits or cetrimide solution to sterilise the mouth piece after each use

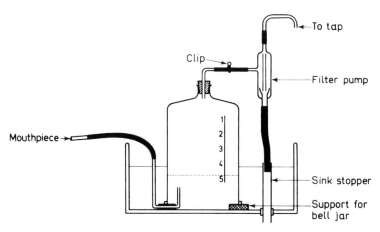

Fig. G5. *Apparatus to demonstrate vital capacity*

*Procedure*

The apparatus is arranged as in the illustration. The sink is filled with enough water so that when the bell jar is full its base remains well covered.

The bell jar is filled to the top mark by using the filter pump. When filled as required the connection to the filter pump should be clipped off.

The subject inhales deeply and then blows through the mouthpiece to empty the lungs as completely as possible without re-inhaling. The bell jar should be held steady during this operation and the inlet tube beneath the bell jar base should not be allowed to tip. The amount of water displaced equals the vital capacity of the individual.

*Further work*

It is interesting to compare the vital capacities of different age groups, sexes, heights and weights, smokers and non-smokers, athletes and non-athletes.

PART H

# Circulation. The Structure and the Properties of Blood

## SECTION I CIRCULATION

### 1. OBSERVATION OF CAPILLARY CIRCULATION

Capillary walls consist of a single layer of cells with a few connective tissue fibres only. The walls are transparent and where the capillaries are close enough to the surface of the skin to be examined under magnification, blood may be seen circulating within these very fine vessels.

#### 1.1. CAPILLARY CIRCULATION IN TROUT ALEVINS OR AMPHIBIAN LARVAE

*Materials required*—Trout alevins/*Xenopus laevis* tadpoles/*Rana* tadpoles/*Triturus* (newt) tadpoles

MS–222 Sandoz (Tricaine methane sulphonate)

Petri dish (40 mm diameter)

Wide mouthed teat pipette

Microscope

*Procedure*

Prepare 1 : 10,000 aqueous solution of MS–222 Sandoz.

One alevin or tadpole should be placed in some of this solution in the Petri dish. Place the dish on the stage of a microscope and wait until the animal is immobilised before trying to focus on part of the capillary network.

The capillaries are most obvious in the thinnest part of the tail fin or in the external gills of tadpoles. If a trout alevin is being studied the capillaries of the yolk sac may also be observed.

Note any differences in the size of blood vessels and compare the rate of flow of blood in the different vessels.

## 1.2. CAPILLARY CIRCULATION IN THE HIND FOOT WEB OF ADULT *XENOPUS LAEVIS* OR *RANA*

*Materials required*—Adult *Xenopus* or *Rana*

MS–222 Sandoz (Tricaine methane sulphonate)

Microscope

Tray (*see Fig. H1*)

Thread

Pipette

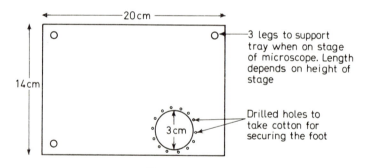

Fig. H1. *Frog tray*

*Procedure*

Prepare a 1 : 2000 aqueous solution of MS–222 Sandoz. Anaesthetise the adult frog (*see* Part D2.4). When immobile place it on the tray with one hind foot spread out as illustrated. Keep the whole body of the frog damp during the observations. Keep the digits in one place by securing them with cotton threaded through the holes in the board (*see Fig. H2*).

Place the tray on the microscope stage so that part of the web is under the objective. Examine the capillary network. Do not let the web dry out, use the pipette to apply water.

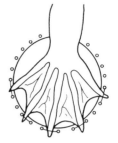

Fig. H2. *Diagram to show the position of foot with web held out secured by thread*

*Further work*

Use an acetylcholine chloride solution (0·05 g per litre) and after thorough rinsing, an adrenaline hydrogen tartrate solution (0·1 g per litre) to irrigate the hind foot web. Observe carefully the rate of blood flow in the capillaries of the web after the application of each solution. Both chemicals are obtainable from:—

> BDH Chemicals Ltd.,
> Poole,
> Dorset.

## 2. INVESTIGATION OF EXERCISE ON PULSE RATE

### The arterial pulse

The measurement of pulse rate is a convenient way of determining the rate of heart beat.

At each contraction of the left ventricle blood is discharged into the aorta. The impact caused by the entry of blood into the arterial system is transmitted as a pulsation along the arterial walls. The pulse resulting from each heart beat travels faster than the blood ejected from the heart at each beat.

The pulse wave velocity is from 5–8 m/s and this is from 10–15 times more rapid than the rate of blood flow.

*Location of radial pulse*

Turn one hand so that the palm faces upwards. Place the fingers of the other hand firmly near the outer border of the arm just below the wrist of the upturned hand. The pulsation of the radial artery should then be felt.

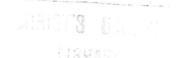

*Requirements for each pair of students—*
                        Chair or stool
                        Watch with second hand or stop watch
                        Graph paper

*Procedure*

(*a*) One student should sit comfortably at rest while the second student records his/her pulse rate.

Count and record the number of beats in successive 15 second periods for 3 min.

(*b*) The first student should then be asked to exercise for 2 min. Where space is limited stepping on and off a chair repeatedly for 2 min should be sufficient.

After 2 min exercise count and record the number of beats in successive 15 second periods continuing until the resting value is obtained.

Plot the graph of the number of beats/15 s (*y* axis) against time in seconds (*x* axis).

*Discussion*

Make a list of the factors which you feel may contribute to a change in the pulse rate during and after exercise.

## 3. AN INDIRECT METHOD OF MEASURING HUMAN ARTERIAL BLOOD PRESSURE USING A SPHYGMOMANOMETER AND A STETHOSCOPE

### Arterial blood pressure

The powerful contractions of the ventricles, particularly the left ventricle, are responsible for pumping blood out of the heart at a relatively high pressure.

A flow of blood through the circulatory system is maintained as blood passes from this region of high hydrostatic pressure to a region of lower pressure.

An adequate arterial blood pressure must be constantly maintained so that sufficient blood is supplied to all the tissues of the body.

If measurements of the arterial blood pressure are made it will be found that the pressure fluctuates from a high level of around 120 mm Hg above atmospheric pressure to a lower level of around 80 mm Hg. These two values may be related to two phases of the cardiac cycle; a phase of ventricular contraction followed by a phase of relaxation.

## Systolic pressure

The maximum pressure recorded. This is the pressure of arterial blood at the height of contraction of the left ventricle.

## Diastolic pressure

This is the lower pressure recorded and is obtained during ventricular relaxation.

These two pressures may be determined with the aid of a sphygmomanometer and a stethoscope.

*Requirements for each pair of students—*
Sphygmomanometer—obtainable from:
Matburn Surgical Equipment Ltd.,
6, Vere Street,
London, W.1
Stethoscope

### DESCRIPTION OF SPHYGMOMANOMETER

The sphygmomanometer consists of two main parts, (*see Fig. H3*). One part consists of a hand pump which is used to increase the air pressure in an inflatable rubber bag. The bag is enclosed in a material case, referred to as the cuff, and is attached by a length of rubber tubing to the second part which is a mercury manometer to measure the pressure of air in the bag.

### THE USE OF THE STETHOSCOPE

A stethoscope is used to listen to sounds produced in the brachial artery distal to the cuff as the cuff pressure is reduced.

You may need practice in the successful application of the stethoscope.

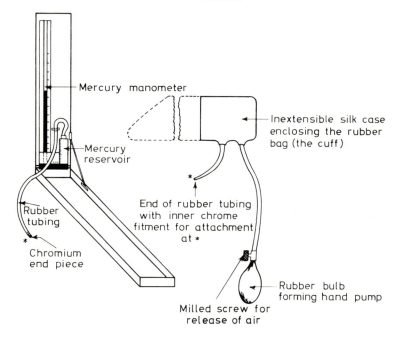

Fig. H3. *The sphygmomanometer*

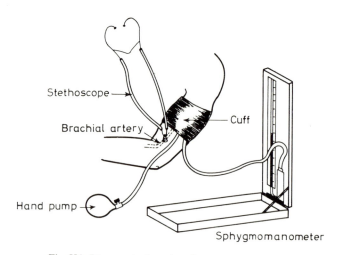

Fig. H4. *Diagram to show the sphygmomanometer in use*

*Procedure for measurement of systolic and diastolic pressure*

Check that there is no reason why the blood pressure measurements should not be made on the student concerned.

The student whose blood pressure is to be measured should be seated with the left arm bared and slightly flexed and with the whole forearm supported at about heart level.

The second student should place the centre of the bag over the brachial artery with the lower edge of the bag one inch above the bend in the elbow.

Wrap the cuff around the bag and the arm tucking in the loose end of the cuff.

With the right hand use the hand pump (with the release screw tightened so preventing the escape of air) to inflate the rubber bag.

With the fingers of the left hand take the radial pulse of the sitting student.

Continue to inflate the bag gently and when the radial pulse disappears note the level of mercury in the manometer. This pressure is approximately the systolic pressure.

Continue to increase the pressure to about 10 mm above this point, *do not apply excessive pressure.*

Apply the stethoscope lightly over the brachial artery at the bend in the elbow (*see Fig. H4*).

The blood flow in the brachial artery will have been obstructed by the inflated bag and no heart sounds will be audible.

Turn the release screw on the hand pump slowly and evenly to allow the gradual escape of air from the compressor bag, at the same time listening for tapping sounds with the stethoscope. When the sounds first appear note the manometer reading. This pressure is taken as the systolic pressure.

Continue to release air at the rate of 3 mm Hg per second thus reducing the pressure applied to the brachial artery. The sounds in the stethoscope become gradually louder, then become muffled finally disappearing altogether. The manometer reading at which the sounds first become muffled is taken as the diastolic pressure.

At this point the cuff pressure is not sufficient to overcome the diastolic blood pressure in the brachial artery and the lumen of the artery remains open.


*Further work*

Measure the arterial blood pressure after 10 min exercise taking readings at 10 min intervals until constant values are obtained.

## 4. EFFECT OF A RISE IN EXTERNAL TEMPERATURE ON THE RATE OF CONTRACTION OF THE DORSAL BLOOD VESSEL OF *NEREIS DIVERSICOLOR*

The dorsal blood vessel is a longitudinal vessel which is visible through the body wall of *Nereis*. The muscular walls of the vessel contract rhythmically to move blood forwards from the posterior end of the animal.

The blood of *Nereis* contains haemoglobin so that the course of the vessel is well marked.

For this exercise it is most convenient to work in pairs with one student responsible for maintaining sea water at the required temperature while the second student times and records the contractions.

*Materials required per pair of students—*

>             2 specimens *Nereis diversicolor* (1 experimental
>                and 1 control)
>             Sea water
>             Small pie dish/crystallising dish
>             Ice
>             1 litre beaker
>             Two 250 cm³ beakers
>             Tripod, gauze, bunsen burner
>             2 thermometers
>             Stopwatch or watch with second hand
>             Glass rod to act as stirrer

*Procedure*

Three-quarters fill one 250 cm³ beaker with sea water and stand the beaker in the 1 litre beaker partly filled with ice, or chill in a refrigerator if available.

Three-quarters fill the second 250 cm³ beaker with sea water and warm it but be careful not to heat the water above 60°C.

Place each worm in a separate pie dish or crystallising dish and just cover with seawater. Label one dish CONTROL.

In each dish place a thermometer to record the temperature of the seawater.

Record the exact temperature of water in each dish.

For each worm count the number of contractions of the dorsal blood vessel in 30 seconds. Make 5 readings. Leave the control dish at room temperature and count contractions at the same time as a count is made for the experimental animal.

Lower the temperature of the sea water in the experimental dish by pouring some chilled sea water in at the side of the dish. Continue pouring and stirring until the temperature of water is about 5°C. Record the exact temperature. Make readings as before. Keep the temperature constant during the readings. Cautiously add a little of the warmed sea water and stir quickly. Aim to raise the temperature by about 5°C. Continue to record at increasingly higher temperatures. If you wish to keep your worm alive, do not raise the temperature of the water above 37°C.

Make recordings in tabular form as shown below:—

| Contractions/30s 5 readings at each temperature | Temp. at start of expt. | Temperature °C | | | | | | |
|---|---|---|---|---|---|---|---|---|
| | | 5° | 10° | 15° | 20° | 25° | 30° | 35° |
| 1 | | | | | | | | |
| 2 | | | | | | | | |
| 3 | | | | | | | | |
| 4 | | | | | | | | |
| 5 | | | | | | | | |
| Average number/30s | | | | | | | | |
| Control worm contractions/ 30s room temperature | | | | | | | | |

Express your results graphically:—
  *x* axis = temperature in °C
  *y* axis = average number of contractions per 30 s
What conclusions are you able to draw from your results?
What was the value of the control worm in this experiment?
Suggest other factors which might affect the rate of contraction of the dorsal blood vessel when *Nereis* is kept in a confined volume of sea water. Suggest ways of investigating the effects of the factors you have listed.

## 5. EFFECT OF CHEMICALS ON THE RATE OF HEART BEAT OF *DAPHNIA*

The heart of *Daphnia* is an oval sac situated dorsally just behind the head (*see Fig. H5*). The contractions of the heart push the pale straw coloured blood out of the anterior end of the heart and into the perivisceral sinuses. Blood re-enters the heart from the surrounding sinus through a pair of laterally placed ostia.

The carapace of *Daphnia* is transparent enabling the contracting heart in lateral view to be easily seen under magnification.

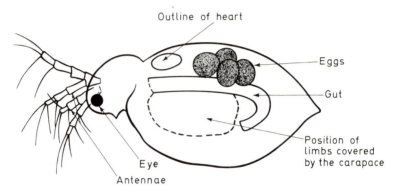

Fig. H5. *Outline diagram to show the position of the heart in* Daphnia

*Materials required—Daphnia* culture
Cavity slides (cavity 15 mm diameter)
Cover slips
Pipettes
Cotton wool
Compound microscope (preferably without built in illumination to avoid heating *Daphnia* and introducing another factor)
Stop watch

*Suggested chemicals—*
Chlorobutol—1 g in 125 cm³ distilled water
Alcohol—5%, 10%, 20%
For further work the effect of caffeine on heart rate might be investigated. Pro-plus tablets, which are available without prescription from chemists, are a source of caffeine. Each tablet contains 50 mg.

RECORDING THE RATE OF HEART BEAT

Place one *Daphnia* in a cavity slide containing a minimum amount of water from the culture container. Add a few cotton fibres to prevent free movement of *Daphnia* and to allow you to observe the rate of heart beat. Compared with the human pulse rate and with the rate of contraction of the dorsal vessel of *Nereis* the heart rate of *Daphnia* is very fast (e.g. 150 beats/min at 19°C).

If one student is working alone to record the rate of heart beat a stop watch is needed. The most convenient method is to have *Daphnia* under observation and hold the stop watch at the same time. Start the stop watch and simultaneously begin to count the heart beats. Count 25 beats and on the 25th beat stop the watch. Record the number of seconds passed during 25 beats.

If students are working in pairs, a watch with a second hand will be adequate and direct recordings of beats/min may be made.

PROCEDURE FOR EXAMINING THE EFFECTS OF CHEMICALS

Make 3 determinations of the heart rate and from these measurements calculate the average heart rate/min. Remove most of the water from the cavity slide by drawing it off with blotting or filter paper held to just touch the edge of the cavity.

Place enough test solution on the cavity slide to just cover the *Daphnia*. Leave the slide for about 3 min then make 3 more measurements of the heart rate.

Use a fresh specimen of *Daphnia* for each test solution, each time recording the heart rate in culture water before applying the test solution.

Record in tabular form as shown below:—

|  | *Specimen 1* | *Specimen 2* | *Specimen 3* | *Specimen 4* |
|---|---|---|---|---|
| *Reading of heart rate* | *Culture water* *Chloro-butol* | *Culture water* *5% alcohol* | *Culture water* *10% alcohol* | *Culture water* *20% alcohol* |
| 1 |  |  |  |  |
| 2 |  |  |  |  |
| 3 |  |  |  |  |
| *Average rate per min* |  |  |  |  |
| *Difference in rate in culture and test solutions* |  |  |  |  |

*Interpretation*

Collect results from other members of the group so that you have
more data to examine before you make any general conclusion about
the effects of the chemicals used.

# SECTION II  STRUCTURE AND PROPERTIES OF BLOOD

For method of examination of blood cells see Part C2.

## 6.  A RED CELL COUNT USING A HAEMOCYTOMETER

The object of this exercise is to make an estimation of the number
of red blood cells per mm³ of human blood. A known quantity of
blood is taken and is diluted to a further known volume with
a diluting fluid. The number of cells in a very small known volume
of diluted blood is counted and the number finally multiplied to give
the number of cells present in a mm³.

*Requirements for each estimation—*

> Improved Neubauer haemocytometer, cover
> slip and graduated pipette (*see Fig. H6*),
> supplied by T. Gerrard & Co. Ltd.
> Disposable sterile lancet
> Cotton wool
> Industrial methylated spirits
> Microscope
> Diluting fluid of the following composition:—
> > 0·6 g sodium chloride
> > 1·0 g sodium citrate
> > 1·0 cm³ 1 % formalin
> > 99·0 cm³ distilled water

DESCRIPTION OF SLIDE AND PIPETTE

The slide is divided into three parts. The surface of the smaller
rectangular area in the centre of the slide is 0·1 mm lower than the
other 2 parts so that when the cover slip is placed over the centre of
the slide a chamber of depth 0·1 mm is formed. The upper surface
of the central region bears a set of finely etched squares (*see Fig. H7*).

When examining the slide use a microscope to locate this ruled area. Each of the smallest squares has a side of length 1/20 mm and thus an area of $1/400$ mm$^2$.

The pipette used to dilute the blood sample is graduated at 0·5, 1·0 and 101 units. The bulb of the pipette has a volume of 100 units, the stem of the pipette a volume of 1 unit.

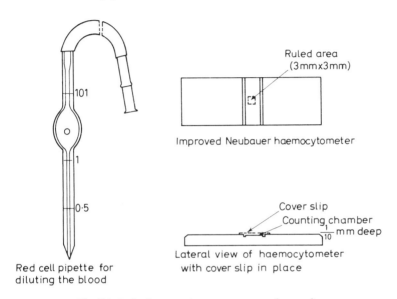

Ruled area
(3mm×3mm)

Improved Neubauer haemocytometer

Cover slip

Counting chamber

$\frac{1}{10}$ mm deep

Lateral view of haemocytometer with cover slip in place

Red cell pipette for diluting the blood

Fig. H6. *Red cell pipette, haemocytometer and cover slip*

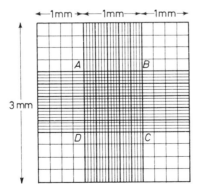

Fig. H7. *Ruled area of the Improved Neubauer haemocytometer.* ABCD *marks the area used for counting erythrocytes. Each heavy line represents a triple ruling*

PROCEDURE FOR OBTAINING A KNOWN VOLUME OF DILUTED BLOOD

Clean the underside of the tip of one finger with cotton wool moistened with industrial meths. Allow to dry. With a sterile lancet prick the finger and obtain a drop of blood. Use the rubber tubing attached to the pipette to suck blood up to the 0·5 mark. (Make certain that you have no air bubbles in your blood sample.) Wipe off any blood from the outside of the pipette.

Suck up diluting fluid to the 101 mark.

Rotate the pipette to mix the blood and diluting fluid. The 0·5 units of blood have been diluted by 99·5 units of diluting fluid, the dilution is 1 in 200.

PROCEDURE FOR FILLING THE COUNTING CHAMBER

Stand the haemocytometer on a piece of filter paper on a flat surface.

Place the cover slip firmly in position over the counting chamber.

Remove the rubber tubing from the pipette and rotate the pipette.

Place the tip of the pipette on a piece of filter paper and draw off the fluid from the stem of the pipette and a small amount from the bulb.

Place the tip of the pipette against the edge of the cover slip. The diluted blood will enter the counting chamber by capillary attraction.

When the counting chamber is full if necessary use a piece of filter paper to blot up any excess fluid outside the cover slip.

The volume of diluted blood over each of the smallest squares in the counting chamber is:—

$$\frac{1}{400} \times \frac{1}{10} = \frac{1}{4000}\,\text{mm}^3$$

PROCEDURE FOR COUNTING THE RED BLOOD CELLS

Place the slide on the microscope stage and use low power to focus on the counting chamber.

Then focus under high power.

Refer to *Fig. H8* and to the slide to see that there are 25 large squares in the ruled area each containing 16 small squares.

Count the red blood cells in 80 of the small squares occupying 5 of the larger squares across the centre of the counting area.

Count all the cells in each square including any which touch the top and left sides of the square but disregard any which touch the bottom or right side of the square. In this way you will avoid counting the same cell twice (*see Fig. H9*).

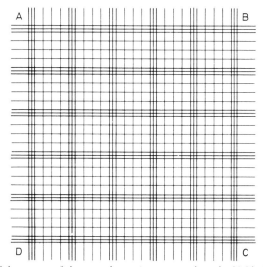

Fig. H8. *Enlargement of the central counting area to show the 25 blocks each of 16 squares each with a side length 1/20 mm*

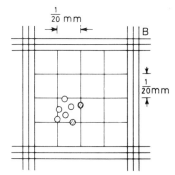

Fig. H9. *Cell counting. Five erythrocytes are counted in this square. The shaded cells touching the bottom and right side of the square are not included for this square*

*To find the total number of cells in 1 mm³ of blood—*

$x$ = the number of cells counted over 80 squares

80 squares represents a volume of $80 \times \dfrac{1}{4000}$

mm³ $= \dfrac{1}{50}$ mm³

There are $x$ cells in $\dfrac{1}{50}$ mm³ of diluted blood

∴ There are $50x$ cells in 1 mm³ of the diluted blood

∴ The number of cells in 1 mm³ of undiluted blood $= 50x \times 200$

$= 10,000x$

## 7. DETERMINATION OF THE HAEMOGLOBIN CONTENT OF THE BLOOD USING THE HALDANE HAEMOGLOBINOMETER

The colour density of the test sample of haemolysed blood is compared with a standard solution whose colour is equivalent to a known amount of haemoglobin per 100 cm³ of blood. The haemoglobin content of the sample is calculated as a percentage of the standard.

Haemoglobin is first converted to carboxy haemoglobin to prevent the formation of the less stable oxyhaemoglobin which might cause variations in the colour of the solution.

*Materials required for each investigation—*

Haldane Haemoglobinometer (*see Fig. H10*), supplied by T. Gerrard & Co. Ltd.
Disposable sterile lancet
Cotton wool
Industrial methylated spirits
Distilled water
Rubber tubing (to fit gas supply tap) with one end attached to a pipette

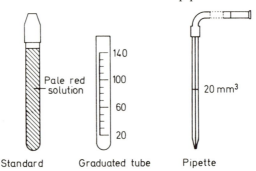

Fig. H10. *Haldane haemoglobinometer*

*Procedure*

Fill the graduated tube up to the 20 mark with distilled water.

Clean the tip of one finger with cotton wool moistened with industrial meths. Allow to dry.

With a sterile lancet obtain a drop of blood from the cleaned finger tip.

Suck up blood to the 20 mm³ mark on the pipette being careful not to include any air bubbles.

Wipe off any blood from the outside of the pipette.

Expel all the blood from the pipette below the distilled water in the graduated tube. Shake the tube well.

Bubble coal gas into the solution of diluted blood to convert the haemoglobin to carboxy haemoglobin. Continue to add distilled water until the colour of the solution in the graduated tube matches the standard.

*Interpretation*

The number on the graduated tube nearest the level of fluid is the % of haemoglobin in the sample

e.g. 80 on the graduated tube means that the amount of haemo-globin in 100 cm³ of the blood under test $= \dfrac{80 \times 14 \cdot 8}{100}$ g

Once the haemoglobin content of the blood has been determined the oxygen carrying power of the blood may be calculated.

1 g haemoglobin will combine with 1·34 cm³ oxygen. If the haemoglobin content is 14·5 g per 100 cm³ blood then every 100 cm³ blood may carry as a maximum 14·5 × 1·34 cm³ oxygen.

*Discussion*

Try to find out if the red blood cell count and the haemoglobin content are affected by environmental and physiological factors.

## 8. BLOOD GROUPING

Human blood may be divided into different types or groups. There are four main blood groups divided from one another by the presence or absence of two **agglutinogens** which are found associated with the red blood cells. The two agglutinogens are termed **A** and **B**.

| Agglutinogens on red blood cell | Blood group |
|:---:|:---:|
| A | A |
| B | B |
| A and B | AB |
| Neither A nor B | O |

In the plasma there may be one, both or neither of two **agglutinins.** These are termed anti-**A** and anti-**B.**

| Agglutinins in plasma | Blood group |
|---|---|
| anti-B | A |
| anti-A | B |
| Neither agglutinin | AB |
| anti-A and anti-B | O |

If anti-A agglutinin is mixed with red blood cells of group A or anti-B agglutinin is mixed with group B blood cells, agglutination or clumping of the red blood cells occurs.

## Rhesus factor

Blood is said to be Rhesus positive if agglutinogen D is associated with the red blood cells and Rhesus negative if D is absent.

The presence of D does not exclude the presence of A and/or B so that any of the 4 main blood groups in the ABO system may be either Rhesus positive or negative.

A Rhesus negative person does not naturally contain anti-D agglutinin in his/her plasma. Anti-D agglutinin is only produced in response to the presence of D. If a Rhesus negative person is accidentally given a first transfusion of Rhesus positive blood there is no agglutination of red cells. Anti-D is slowly produced in response and if a further transfusion of Rhesus positive blood is given, this time the red cells will be clumped by the anti-D agglutinin.

## 8.1.  BLOOD GROUPING USING TWO TEST SERA FOR THE ABO SYSTEM

*Materials required*—Anti-A blood grouping serum
Anti-B blood grouping serum
obtainable from:—
A. R. Horwell (Reagents) Ltd.,
Laboratory and clinical supplies,
2 Grangeway Kilburn High Road,
London, N.W.6.

or:—
> Baxter Laboratories Ltd.,
> Caxton Way,
> Thetford,
> Norfolk

Microscope slides
Tooth picks
Wax pencil
Sterile lancets
Cotton wool
Industrial methylated spirits

*Procedure*

Use the wax pencil to divide the surface of a clean microscope slide into two parts.

Place a drop of anti-A serum in the left space and a drop of anti-B serum in the right.

| Anti-A | Anti-B |
|:---:|:---:|

Clean a finger tip with cotton wool moistened in industrial methylated spirits. Allow to dry. Use a sterile lancet to obtain a drop of blood from the finger.

Use separate tooth picks to transfer a small drop of blood to each of the test sera. Mix the blood and serum but be very careful not to allow one serum to run into the other.

If clumping of the red blood cells occurs it will occur a few seconds after the blood and serum are mixed. Examine the slide under the low power of a microscope.

Agglutinated cells look like small red grains in a clear liquid.

If there is no agglutination the fluid remains uniformly pink.

*Interpretation*

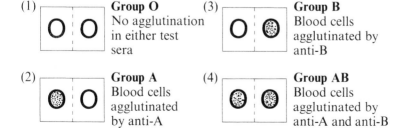

(1) **Group O** No agglutination in either test sera

(2) **Group A** Blood cells agglutinated by anti-A

(3) **Group B** Blood cells agglutinated by anti-B

(4) **Group AB** Blood cells agglutinated by anti-A and anti-B

## 8.2. BLOOD GROUPING USING ELDON CARDS FOR THE ABO AND RHESUS SYSTEMS

*Materials required*—Eldon cards (*see Fig. H11*). The cards are supplied with pipette, plastic sticks and instructions for use. Obtainable from Philip Harris Ltd.

> Sterile lancets
> Industrial methylated spirits
> Cotton wool
> Beaker of water

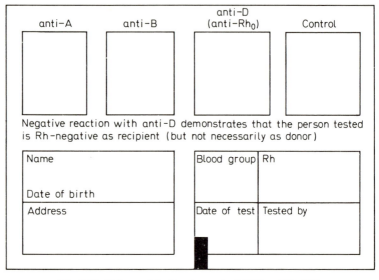

Fig. H11. *Eldon card for blood grouping* [*ABO and Rh$_o$ (D)*]

*Procedure*

(*a*) Use the pipette supplied in the pack to add one drop of tap water to each test panel.

(*b*) With the flat end of one of the plastic sticks provided mix the water and reagent in the centre of the test panel (labelled anti-A).
  Clean the stick with cotton wool.
  Repeat stage (*b*) for the other three panels.

(*c*) Clean the tip of a finger with cotton wool moistened with methylated spirits. Allow to dry. Obtain a drop of blood by pricking the finger with a sterile lancet.

(*d*) Transfer some of the blood from the finger on to the flat end of a plastic stick. There should be enough blood to cover the end of the stick but not enough to overflow down the sides. Mix this blood with the dissolved reagent in panel anti-A. When mixed spread the blood and reagent over the whole panel. Clean the plastic stick.

Repeat stage (*d*) for each of the three remaining panels.

Wait one minute.

(*e*) Thoroughly mix the blood and reagents in each panel by holding the card vertically facing in each of 4 different directions for 10 seconds.

Check that the contents of one panel do not run into the next panel.

Repeat stage (*e*) twice more.

Note the appearance of the test mixture on each panel.

*Interpretation*

A positive reaction between blood and reagent results in agglutination, the red blood cells are clumped and the mixture has a granular appearance.

If there is no agglutination the test mixture appears uniformly red and non granular. The control panel should always have this appearance.

Group O is not agglutinated by either anti-A or anti-B.

Group A is agglutinated only by anti-A.

Group B is agglutinated only by anti-B.

Group AB is agglutinated by both anti-A and anti-B.

Rhesus positive is agglutinated by anti-D.

Rhesus negative is not agglutinated by anti-D.

*Discussion*

(1) Having seen the results of the reaction between agglutinin and agglutinogen you should be aware how necessary it is to know the correct blood group of a patient before any blood transfusion is contemplated.

To be completely safe blood of the same group as that of the patient should be given but if this is not available it is essential that a compatible group is used.

With a limited transfusion the plasma of the donor has a negligible effect on the red blood cells of the recipient as the recipient's plasma will dilute the donor plasma to a level where the agglutinis will be ineffective. The important point to remember is that the red cells of the donor must not be agglutinated by the plasma of the recipient.

Complete the table below to show the results of possible transfusions.

+ = agglutination
− = no agglutination ( ∴ transfusion possible)

*Blood group of recipient*

|  |  | A | B | AB | O |
|---|---|---|---|---|---|
| *Blood group of donor* | A | − |  |  |  |
|  | B | + |  |  |  |
|  | AB | + |  |  |  |
|  | O | − |  |  |  |

From your complete table you should be able to see which group is described as the **universal recipient** and which group is described as the **universal donor.**

(2)  Record the number of individuals of each blood group in your class. Calculate the percentage frequencies of each group.

If your class frequencies differ from the percentage frequencies for the U.K. given here use the chi-squared test (*see Part E2*) to see if the differences are significant or not.

PERCENTAGE FREQUENCY OF THE ABO BLOOD GROUPS

(*Examples obtained from 'The ABO Blood Groups' by A. E. Mourant* et al. *by courtesy of Blackwell Scientific Publications*)

|  | O | A | B | AB |
|---|---|---|---|---|
| *UK (RAF personnel)* | 46·68 | 41·72 | 8·56 | 3·04 |
| *Russians (Leningrad)* | 31·96 | 33·92 | 26·08 | 8·04 |
| *Lapps (Sweden)* | 28·96 | 62·62 | 4·46 | 3·96 |
| *Aborigines (Kimberley, W. Australia)* | 60·70 | 39·30 | 0·00 | 0·00 |
| *Indians (Argentine)* | 96·30 | 3·70 | 0·00 | 0·00 |

## 9. THE CLOTTING OF BLOOD

*Materials required*—Sterile lancets
          Industrial methylated spirit
          Cotton wool
          Microscope slides
          Microscope
          10% sodium citrate
          1% potassium oxalate
          5% calcium chloride
          Fine glass capillary tubing
          Stop watch

*Procedure*

(1) Use a sterile lancet to obtain a few drops of blood from a previously cleaned finger. Place the blood on a microscope slide and examine under a microscope. Note the time taken for the blood to clot.

(2)(*a*) Place a few drops of either 1% potassium oxalate or 10% sodium citrate on a clean microscope slide. Add a drop of blood and examine after the time noted in (1) for a clot to form.

(*b*) Add a few drops of calcium chloride solution to the blood on the slide.

Try to explain the results you obtain for (*a*) and (*b*).

(3) Use a sterile lancet to prick a finger. Hold a length of fine glass capillary tube horizontally so that blood from the finger flows up the tube. The stop watch should be started immediately blood enters the tube.

Break off short lengths of the tube every 20 seconds until a thread of clotted blood is pulled from the tube.

Stop the watch and note the time taken for the blood to clot.

If you record the clotting times of other members of a large class you will probably find there is a continuous variation in the results.

A probability curve will be obtained if you plot clotting times against the number of individuals registering each particular time.

# Locomotion

Most animals need to move from place to place in their search for continued supplies of food. A mobile animal is able to explore new areas and when necessary may search for a mate and a suitable place to establish a home territory. An animal which is able to move may escape from conditions which are unfavourable to it, perhaps a part of the environment which has become overcrowded, or from the presence of a predator.

In acellular organisms cilia and flagella may be used to effect movement but the forces set up by these organelles are small and are incapable of producing movement in most animals larger than the protozoans, some metazoan gametes, larvae or small platyhelminths.

Coelenterates possess musculo-epithelial cells which are differentiated for contraction and these cells may have quite a complex organisation as in the sea anemone, *Metridium*. The type of cell arrangement found in *Hydra* is relatively simple.

Above the coelenterates locomotion is accomplished by the contraction of specialised muscle tissue acting together with skeletal structures. Both unstriated and striated muscle are concerned with movement.

Compared with the lower invertebrates the more advanced forms have a greater proportion of striated muscle and this can be correlated with an increase in activity.

In vertebrates the jointed skeleton is moved by the contraction of striated muscle only. The muscles are arranged in antagonistic pairs, a contraction of one member of the pair bringing about an opposite effect to that caused by a contraction of the other muscle.

A muscle cell only contracts on the receipt of a stimulus and in the intact animal this stimulus is in the form of a nervous impulse.

In isolated nerve-muscle preparations in the laboratory muscles may be made to contract as a result of the application of electrical, chemical or mechanical stimuli.

When a muscle is stimulated energy for contraction is released from the breakdown of molecules of A.T.P. within the muscle cells.

## 1. MOVEMENT IN THE PROTOZOA

Protozoans move in a variety of ways, some species employing more than one method of locomotion. Some move as a result of reversible changes which occur in the general cell cytoplasm. Movement which results from cytoplasmic streaming, with the formation of pseudopodia, is known as amoeboid movement. In some organisms parts of the cytoplasm have become differentiated to form fibrils (myonemes) and it seems likely that these are concerned with movement. Fibrils are found in the ciliates with contractile stalks, they are also present in the cytoplasm of sporozoans where they may bring about the gliding movement described as gregarine movement.

Some protozoans possess locomotory organelles clearly differentiated from the rest of the organism, examples are the flagella and cilia. Electron microscopy has shown that both flagella and cilia have the same fundamental structure, the difference between the two being mainly in their length and distribution. Flagella are normally longer than cilia and they are developed singly or in small numbers whereas the shorter cilia are found in masses, sometimes distributed over the whole surface of the organism, their beat being co-ordinated to show a metachronal rhythm.

INTRODUCTION TO PRACTICAL WORK

A microscope which provides × 400 magnification is essential.

Dark field illumination should be tried. To achieve a dark field different sized discs of black paper on clear glass should be placed just beneath the substage condenser. The larger discs being used with the higher power objectives.

Phase contrast equipment is very valuable for the examination of non-stained transparent Protozoa. With this type of equipment small differences in the refractive indices of the cytoplasmic components are shown as differences in brightness of the image.

Philip Harris Ltd. stock a Watson Student Phase Unit which fits all Watson and most other British made instruments. This unit includes a × 40 phase contrast objective.

Before you begin your observations you may need to prepare some microscope slides. For examination under high power the drop of culture medium must be covered with a cover slip. Various substances may be used to support the cover slip but to be satisfactory they should serve the double function of preventing the organism being squashed and preventing the evaporation of the drop of liquid. Vaseline serves these purposes.

Apply to the surface of a slide a thin ring of vaseline the same shape or a little bit smaller than the cover slip to be used. The drop of culture should then be placed in the centre of the area marked out by the vaseline and when the cover slip is placed in position it will be supported and the liquid will be contained in a sealed compartment.

To avoid wasting time looking for organisms in a drop, the culture may be concentrated beforehand using a hand centrifuge before a drop is extracted.

*Substances used to slow down movement—*

Methyl cellulose—Add 2 g to 98 cm³ water 24 h before required. Use 1 drop with 1 drop of culture. You may need a higher concentration for ciliates

Sodium alginate (Manucol)—Start with 0·1% solution and if necessary adjust the concentration for your requirements.

*Routine procedure for studying Protozoa*

(1) Examine a drop of the culture first under low power and then under high power using transmitted light.

(2) Repeat your observations but with a dark ground illumination.

(3) Use phase contrast equipment if available. For good results you will need to gently compress the protozoan before observation.

(4) If movement is too rapid for you to observe anything but the path traced by the animal use one of the suggested chemicals to slow the organism.

(5) If necessary use a vital stain e.g. neutral red to show additional structural details.

SELECTED PROTOZOA FOR STUDY

*Amoeba*⎫
*Euglena*  ⎬ Cultures obtainable from the
*Paramecium*⎭    main biological suppliers
*Vorticella,* or other stalked ciliates, obtainable
    from pond water
*Opalina,* from the rectum of a frog
*Monocystis,* from the seminal vesicles of an
    earthworm

*Materials required for all observations—*
Microscope
Thin glass slides
Cover slips
Vaseline
Teat pipettes
Watch glasses
Methyl cellulose or sodium alginate

## 1.1. *AMOEBA*

*Additional materials required*—Neutral red

*Procedure*

Work through the routine procedure.

Note the streaming movements in the cytoplasm and write a description of what you are able to see.

Make drawings at regular intervals to show the change in shape of an *Amoeba* with the development of different pseudopodia.

Add a drop of $0.01\%$ neutral red solution to the drop of culture solution. Leave for 1 h and examine after this time.

## 1.2. *EUGLENA*

*Procedure*

Work through the routine procedure.

Write a short description of the type of movements taking place.

The position of the moving flagella is marked by movement in the water surrounding it.

With *Euglena* under high power make drawings to illustrate euglenoid movement.

*To fix and show flagella (and cilia)*

Add a drop of Noland's solution to a drop of culture.

*Noland's solution*—80 cm³ saturated phenol in distilled water
20 cm³ formalin (40%)
4 cm³ glycerol
20 mg gentian violet

## 1.3. *PARAMECIUM*

*Procedure*

Work through the routine procedure.

Note the spiral path traced by the organism as it moves.

You should also be able to see that the animal rotates about its longitudinal axis as it travels.

Note which end of the animal leads during forward movement.

INVESTIGATION OF THE RATE OF MOVEMENT OF *PARAMECIUM*

*Additional materials required*—
Eye piece micrometer. This is a glass disc bearing an engraved scale of 100 equal divisions
Micrometer slide. This glass slide bears an engraved scale of length 1 mm subdivided into 0·01 mm divisions
Stop watch

*Preparation*

The eye piece scale must be calibrated against the micrometer slide for *each* objective used.

With the eye piece micrometer in place and the micrometer slide secured to the stage bring the scale on the slide into focus. Move the slide about until the initial division mark on the eye piece scale is superimposed on the initial division mark of the slide. Count the number of divisions on each scale until two other division marks are superimposed.

Each division on the stage micrometer scale $= 10$ μm. Thus the value in μm of one division on the eye piece scale

$$= \frac{\text{number of divisions on micrometer slide} \times 10}{\text{number of divisions on eye piece scale}}$$

*Procedure*

The length of different species of *Paramecium* is from about 100 μm to 300 μm.

Having found the value in micrometres of one division on the eye piece scale you should be able to estimate the approximate length of the species you are observing.

The approximate rate of movement may be found by counting the number of divisions crossed by the animal in a known period of time.

You will need to make a number of readings to give you an idea of the average speed. Record your results in micrometres per second.

You could then investigate the effect of different concentrations of methyl cellulose or sodium alginate on the rate of movement.

## 1.4. FURTHER WORK WITH OTHER CILIATES

Other ciliates should be examined as time permits. Some of the stalked ciliates are commonly found attached to either pieces of plant material or other animals in fresh water. The stalk of *Vorticella* may be seen to contract and the action of the cilia may be investigated by adding a suspension of carmine particles to the water.

Frogs and toads are good sources of protozoa. Remove the rectum from a freshly killed frog and wash the contents into a watch glass containing 0·65% saline.

Examine drops of the solution under high power. *Opalina* and two ciliates, *Nyctotherus* and *Balantidium* may be found.

The intestine and body cavity of *Periplaneta americana* should be examined for parasitic protozoa. Species from each of the four classes of protozoans have been recorded.

## 1.5. *MONOCYSTIS*

Several species of *Monocystis* occur as parasites of earthworms.
Species of *Monocystis* have been recorded from:
> *Allolobophora caliginos*
> *Allolobophora longa*

*Eisenia foetida*
*Lumbricus castaneus*
*Lumbricus rubellus*
*Lumbricus terrestris*

*Monocystis* may be found in the seminal vesicles, the coelom or intestine.

The only stage of *Monocystis* which you are likely to see showing any signs of movement is the trophozoite.

*Materials required*—Earthworm
70% alcohol
0·65% saline
Dissecting dish
Dissecting instruments
2 beakers (250 cm³)
Watch glass

*Procedure*

Kill the earthworm by immersing it in 70% alcohol for a few seconds.

When movement has ceased remove the worm and wash it immediately in 0·65% saline.

Open the worm in a dry dissecting dish. Make a dorsal incision and expose the seminal vesicles.

Remove part of one of the seminal vesicles. Tease out the contents in a watch glass containing enough saline to prevent drying.

Transfer some of the teased contents to a slide making a thin smear on the surface. Add a drop of saline when necessary. Do not let the smear dry up.

Examine under L.P. and H.P., look for trophozoites and watch for any signs of movement.

## 2. MOVEMENT OF THE EARTHWORM

The earthworm is divided into a number of segments separated internally by partitions known as septa. The body wall of each segment consists mainly of an outer layer of circular muscles and an inner layer of longitudinal muscles. Each segment contains incompressible coelomic fluid which acts as a hydrostatic skeleton.

The pressure changes exerted by the coelomic fluid as a result of the contraction of the surrounding muscle fibres are restricted by the septa.

The muscle layers of the body wall are innervated by segmentally arranged peripheral nerves which leave the longitudinal ventral nerve cord. The muscle layers act in an antagonistic way. Contraction of the circular muscles causes a decrease in the diameter and due to the incompressibility of the fluid an increase in the length of a segment while contraction of the longitudinal muscle causes a relative increase in diameter and decrease in the length of a segment.

When an earthworm moves forward the sequence of events is as follows. Circular muscles in the anterior segments contract, the chaetae of these segments are withdrawn and elongation or forward movement of these segments results. Contraction of the circular muscles with a withdrawal of chaetae passes posteriorly. Contraction of the longitudinal muscles in the anterior segments with a protrusion of chaetae then occurs. At those points where the longitudinal muscle is contracted the worm is able to exert a thrust against the substratum. The wave of longitudinal muscle contraction follows the passage of circular muscle contraction and these events continue in a regular alternate way.

*Materials required*—2 earthworms
                 Ethyl alcohol (or industrial methylated spirit)
                 $0.65\%$ saline
                 Wax bottomed dissecting dish
                 Beaker
                 Dissecting instruments
                 Pins

*Procedure*

Observe an earthworm moving in a dissecting dish. Keep the external surface of the worm moist with $0.65\%$ NaCl. Note the changes in the shape of the segments in different regions of the body during forward movement. Make outline diagrams to show the sequence of events which occur.

Place the worm on a sheet of glass and then on a sheet of coarse textured paper. Account for any differences in movement on the two surfaces.

SEGMENTAL CO-ORDINATION IN THE EARTHWORM

These exercises require some dissection of anaesthetised earthworms, they may be more suitable as demonstrations by a member of staff rather than exercises performed by individual students.

(1) Place an earthworm in tapwater in a beaker and gradually add alcohol, stirring all the time, until the worm is bathed in 10% (by volume) alcohol. When the worm ceases to move it will be anaesthetised.

Place the worm in a wax bottomed dissecting dish. Make a dorsal incision, preferably in the intestinal region for ease of dissection, and continue the cut through the body wall of a few segments. Turn back the flaps of the body wall and cover the exposed region with 0·65% NaCl. Expose the ventral nerve cord by cutting away all the surrounding tissues. Pin out the flaps of the body wall (*see Fig. I.1*).

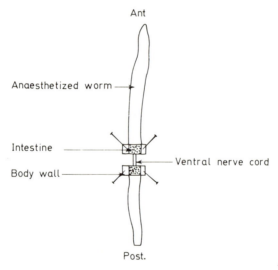

Ant

Anaesthetized worm

Intestine

Body wall

Ventral nerve cord

Post.

Fig. I.1. *Diagram to show the appearance of worm after the exposure of part of the nerve cord*

As the worm recovers watch the waves of activity spread posteriorly. Does the part of the worm posterior to the dissected region begin to move independently of the anterior region or is the movement of the two parts co-ordinated?

If contraction in an anterior segment is followed by contraction in a posterior segment describe from your observations how the two segments may be co-ordinated.

(2) Anaesthetise a second earthworm in the way described previously.

With fine pointed scissors make a cut through the ventral body wall
to sever the ventral nerve cord.

As the worm recovers from the anaesthetic watch carefully to see if
the waves of activity are impeded at the point of the severed nerve
cord.

If there is no interference in the waves of activity how do you think
co-ordination may be effected between two segments which are no
longer linked by the ventral nerve cord?

## 3. MOVEMENT IN *NEREIS*

*Materials required*—1 preserved *Nereis*
                    1 temporary or permanent preparation of a
                        parapodium
                    1 live *Nereis*
                    Large dissecting dish or other suitable container
                    Sea water
                    Microscope

*Procedure*

(1) Obtain a preserved specimen and make a careful examination of
the external features.

The structure of a parapodium should be investigated.

Either examine a permanent preparation or if the experimental
animal is to be killed at the end of the exercise make a temporary
preparation by removing a parapodium from the mid body region
and mounting it in glycerine.

Examine under a microscope and make drawings to show the
structure. How many kinds of chaetae are there?

(2) Carefully watch the processes of crawling and swimming.

Note particularly the action of the parapodia and their co-
ordinated pattern of activity along the length of the body. How is
the structure of a parapodium adapted for the functions it serves?

Do you observe a greater amplitude of the body waves in
swimming?

Place an obstruction in front of *Nereis* when it is moving and
observe what happens.

Investigate the effect which follows the transaction of the ventral
nerve cord of an anaesthetised *Nereis*.

## 4. MOVEMENT IN INSECTS

In adult insects the three segments of the thorax bear the locomotory organs. Each thoracic segment carries a pair of jointed limbs and in addition the mesothorax and metathorax may each have a pair of wings.

The leg of an insect consists of 5 main segments with the free end of the leg terminating in a pair of claws.

The legs show many adaptations for different methods of locomotion. One or more of the three pairs of limbs may be highly specialised for a particular function. In different species the legs may be used for walking, running, jumping or digging and in aquatic insects the legs may show structural modifications for swimming.

In some larval and adult forms there are additional aids to locomotion. In the caterpillars of the Lepidoptera there may be 5 pairs of pro-legs on the abdomen. In the Collembola there is a forked springing appendage on the 4th abdominal segment.

The majority of insects possess wings but some of the primitive orders are wingless and there are others e.g. the ectoparasites whose lack of wings is a secondary modification.

It is common among winged forms to find the 2 wings on one side coupled together and functioning as a single unit. In other orders one pair of wings only may be adapted for flight while the other pair are either specialised for a protective function or are very reduced in form.

SOME INSECTS WHOSE MOVEMENT MAY BE STUDIED IN THE LABORATORY

> *Carausius morosus* (stick insect) *see* 4.2
> *Locusta*
> *Periplaneta*
> *Dytiscus*
> *Notonecta*
> *Aeschna*

### 4.1. GENERAL PROCEDURE FOR THE STUDY OF MOVEMENT

Observe movement of the insects in glass sided laboratory containers where the artificial conditions are as near to the natural conditions as is possible.

Write a short description of the method of movement you observe.

Examine individual limbs from preserved specimens of the species you have been watching. Make outline drawings to show the proportion of each segment of the limb. Try to relate structure and function.

When the limbs are large enough, for example the hind limb of *Locusta*, investigate the musculature by dissection.

Most insects are able to move quite rapidly and thus it is quite difficult to examine the patterns of activity. One insect whose method of progression is a rather deliberate amble is the stick insect.

## 4.2. EXAMINATION OF THE WALKING PATTERN IN STICK INSECTS

*Materials required*—Culture of *Carausius morosus* (obtainable from Philip Harris)
Thread

When an insect walks or runs the six legs are usually divided into two functional groups, each of three legs. Each group consists of the prothoracic and metathoracic limbs on one side together with the mesothoracic limb of the other side.

The three legs of one group are raised from the ground and moved forward together while the other three legs are left on the ground to support the insect in a stable position. These are then raised and advanced leaving the other group as a support.

*Procedure*

Place a stick insect, with all six legs intact, on the bench and observe the alternate movements of each group of three legs.

In a culture of stick insects there are usually some animals lacking one or more legs. You should observe the walking pattern of insects which have legs missing in different positions. If all your insects have a full set of legs try tying individual legs to the side of the body so that each one is temporarily out of action. At first isolate only one leg and then try the effect of isolating two legs, one on each side of the body.

Each time note the grouping of the legs during walking. Is the original grouping altered so that stability is maintained?

*Further work*

Because of the speed with which they occur wing movements are more difficult to distinguish with the unaided eye.

Observations of the wing movements of different insects may be made using a stroboscope (*see* reference list).

## 5. THE GASTROCNEMIUS MUSCLE-NERVE PREPARATION

The reaction of a muscle when its motor nerve is stimulated may be studied by arranging a nerve-muscle preparation in conjunction with a recording kymograph. The isolated muscle (e.g. frog gastrocnemius muscle) with its motor nerve still attached is fixed at one end the free end being connected to a lever. The lever is balanced so that it will pivot freely. A sharp pointer or pen device is fitted to the end of the lever opposite to that connected to the muscle. Movements of the lever cause the pointer to produce a trace on a paper which covers a revolving drum.

Although fresh muscle in isolation may be seen to contract when its nerve is stimulated the recording device gives an appreciation of the magnitude of the contraction adding the dimension of time. A permanent record of the contraction is also produced. Reaction to different rates and strengths of stimuli can be recorded by this method.

The nerve is stimulated by an electric impulse delivered through a pair of electrodes from a source of direct current. The nerve and muscle are usually mounted on a platform or in a special muscle bath.

### DESCRIPTION OF APPARATUS

Complete sets of apparatus suitable for the experiment can be purchased. An advanced set may be relatively expensive. C. F. Palmer Ltd., Carlisle Road, The Hyde, London, N.W.9 produces a muscle holder with a kymograph and stimulator grouped specially for student work (*see Plate I.1*). This supplier also offers all the associated accessories.

However, each item of the entire apparatus can be constructed from reasonably simple materials and will be described separately. Such a grouping may not have the ease of operation or flexibility afforded by the commercial apparatus.

### 5.1. KYMOGRAPH

This is essentially a clockwork or electric motor turning a drum. The speed can be fixed or the motor can be fitted with reduction gears or a speed control. Being able to produce various speeds will enhance the range of experimental work.

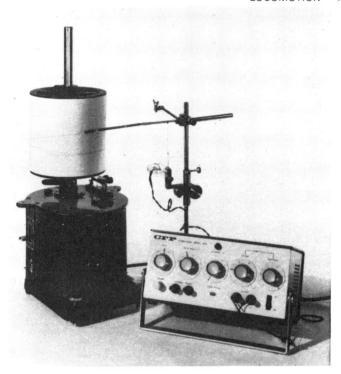

Plate I.1. *Student kymograph, muscle plate, lever and stimulator (Courtesy of C. F. Palmer)*

D. Ryman (1964, Sch.Sci.Rev. 156, 390) has described the con-struction of a versatile and reliable kymograph which employs pulley devices attached to an electric gramophone motor to drive the drum (*see Plate I.2*). The motor is fitted into a wooden frame. By using the pulley wheel dimensions given, the speed of the drum can be varied from 42 rpm to one-third rpm with a motor which rotates at 90 rpm. The driving belts can be rubber or string bands, or the solid tubing used on the pulley system of a dentist's drill. This latter material can be carefully measured to the required size and the ends fused by heating to produce a flexible band.

If a slow speed motor is available a direct connection may be made via the driving band to the drum.

The drum can be made from a large coffee or sweet tin. It must be very carefully drilled for centre at each end and mounted on a spindle. Care in finding the exact centre will be repaid by undistorted traces during an experiment. The drum must be easily removable

Plate I.2. *Kymograph (after D. Ryman)*

from its spindle. A simple arrangement of drum and motor is shown in *Fig. 1.2.*

The trace is obtained on a paper which is tightly mounted on the drum. It is usual to smoke the paper so that an even covering of soot is deposited. This can be done by rotating the paper covered drum over the flame from a fish tail bunsen to obtain an even black/brown deposit. The gas feeding the bunsen should be bubbled through benzene. The operation is best carried out with care in a fume cupboard. Several interchangeable drums can be prepared at one time. Special glazed paper may be purchased as can the smoking apparatus but, glossy paper such as the master paper used for duplicating or shiny lining paper will be suitable.

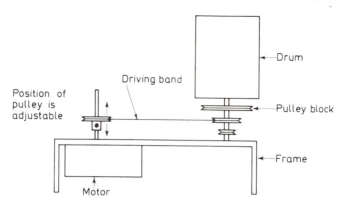

Fig. I.2. *Simple kymograph*

When the experiment is completed the paper can be removed from the drum and after making notations such as name, date, type of stimulation, it should be passed through varnish in a tray or sprayed with varnish and when dry stored for reference.

If an inking device is available such as that used on a barograph it will not be necessary to smoke the paper. If this method is adopted the paper should still be as smooth as possible and it will be necessary to use special barograph ink.

## 5.2. THE LEVER MECHANISM

The lever can be a thin metal or wooden rod. An umbrella rib will act very well as a lever. The position of the pivoting point will determine the magnification of the movement of the contracting muscle. It should be so balanced horizontally that it will easily support the muscle (which in the experiment described is arranged in a vertical position) and, return to the horizontal after contraction.

A simple lever can be made from a plastic drinking straw. This should be pierced with a glass headed pin on which it pivots. The pin is then pushed into a cork. The pin must not hold the straw too tightly but the possibility of side to side movement should be eliminated by inserting plastic washers on either side of the straw (*see Fig. I.3*).

The tracing nib at the end of the lever can be cut from the plastic top of a coffee or baby food tin and inserted into the straw and glued into position (*see Fig. I.4*).

Fig. I.3. *Lever pivot*                    Fig. I.4. *Tracing nib of lever*

## 5.3. THE MUSCLE MOUNT

A muscle chamber made of perspex, with electrodes attached, may be purchased. A more simple arrangement may be constructed with the two electrodes (preferably platinum or graphite *not* copper), pinned to a large cork. The upper surface of the cork should be covered with several layers of filter paper to absorb excess frog Ringer when it is applied to the nerve-muscle preparation (*see Fig. I.5*). The cork is held tightly in a clamp.

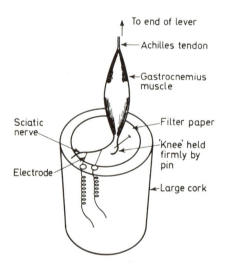

Fig. I.5. *Simple muscle mount*

## 5.4. THE STIMULATOR

An electronic stimulator can be obtained from a supplier.

Current from a 6 V battery will stimulate a fresh frog muscle to contract. Ideally though, there is a need to produce a controlled pulse of current of just the right intensity. An induction coil is normally employed to produce the stimulation.

The induction coil consists of two coils of wire called the primary and the secondary coils. The secondary coil has a hollow core and can be moved over the primary coil. The primary receives its current from a dry cell or 2 V accumulator, a switch or key being included in the circuit (*see Fig. I.6*).

When the current enters the primary circuit it takes a little time to reach its maximum owing to the inductance of the primary coil.

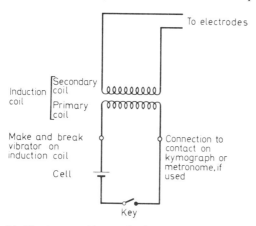

Fig. I.6. *Circuit to provide stimulus for nerve-muscle preparation*

A current is gradually induced in the secondary coil, being at its greatest at the point when the primary circuit is broken. The system is adjusted so that the muscle only contracts at such a break (*see Fig. I.7*). The secondary coil may be moved in relation to the primary coil and this will affect the characteristic of the pulse produced in the secondary coil. It will be necessary to adjust the position of the secondary coil when the experiment is set up in order to achieve the correct impulse for the maximum contraction of the muscle.

Alternative ways of activating the coil may be used other than by making and breaking the primary circuit manually with a key.

(*a*) The primary coil can be wound round an iron core so that when a current passes through the primary circuit the iron becomes magnetised and can be employed to attract a soft iron plate. The plate is attached to a spring and this forms a make and break vibrator included in the primary circuit. As the iron plate is attracted to the iron core the vibrator contacts separate and the circuit being broken the iron core becomes demagnetised. The iron plate then springs back, the contacts touch and the circuit is again completed. The distance between the contacts can be adjusted to alter the speed at which the vibrator operates to open and close the primary circuit

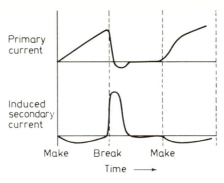

Primary current

Induced secondary current

Make     Break     Make

Time ⟶

Fig. 1.7. *Characteristics of induction coil*

which thus regulates the timing of the secondary coil output. This arrangement gives repetitive pulses.

(*b*) A make and break contact as part of the primary circuit can be attached to the kymograph. The speed at which the drum turns will affect the rate at which this switch operates and the stimuli are delivered to the nerve. One stimulus is applied for each revolution of the drum.

(*c*) A metronome adapted as an adjustable interrupter may be used to give controlled impulses up to 200 contacts per minute.

A tuning fork may be arranged to produce time tracings on the smoked paper. A 100 Hz tuning fork will produce 1/100th of a second time tracings.

DISSECTION OF THE FROG TO REMOVE THE GASTROCNEMIUS MUSCLE AND SCIATIC NERVE

*Materials required*—Pithed frog
                     Dissecting dish and cork mat
                     Scissors
                     Forceps
                     Mounted needle
                     Seeker
                     Thin round ended glass rod
                     Petri dish
                     Rubber teat pipette
                     Frog Ringer—Sodium chloride 1·30 g
                                     Potassium chloride 0·04 g
                                     Sodium bicarbonate 0·04 g
                                     Calcium chloride 0·06 g
                                     (anhydrous)
                                     Distilled water 200 cm$^3$

*Procedure*

The pithing of the frog should be carried out by a competent person.

With the ventral surface of the frog held uppermost grasp the hind feet and kill the frog by rapidly hitting it against a firm flat surface. On the dorsal side cut through the skin of the head and the occipito-atlantal membrane between the skull and spinal column. Insert a needle into the cranial cavity and move it about to destroy the brain. The needle is then pushed down the vertebral canal to destroy the spinal cord.

Place the frog ventral side up in a dissecting dish on the cork mat. Have a good supply of frog Ringer at hand including a full Petri dish into which the muscle preparation should be placed as soon as it is dissected out.

With the forceps grip the skin of the lower abdomen and use scissors to make incisions as shown in the diagram (*see Fig. I.8*). Turn

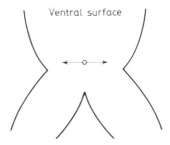

Fig. I.8. *Diagram to show position of first cuts in dissecting out nerve-muscle preparation of frog*

the frog over and continue the cuts to meet in the middle at the back. Peel the skin back over each leg.

Using the fingers carefully push aside the flexor and extensor muscles of the right upper leg (*see A in Fig. I.9*). The larger sciatic nerve will be seen at the proximal end of the leg. It may be hidden beneath the ilio-fibularis muscle which should be removed at this stage. Push back the thigh muscles away from the sciatic nerve using the glass rod dipped in Ringer. Do not forget to keep the now isolated nerve moist with Ringer.

Carefully remove the muscles of the upper leg from the knee joint to ilium. Cut through the urostyle and the transverse processes of the sacrum (*see cut 1 in Fig. I.9*) then cut through the vertebral column above the sciatic complex (*cut 2*). If it is intended to make two preparations using the gastrocnemius of the left leg as well, cut

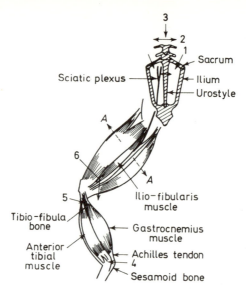

Fig. I.9. *Diagram to show muscle arrangement of right rear leg of frog*

up the vertebral column as indicated by *cut 3* in the diagram (*see Fig. I.9*).

With the glass rod or seeker carefully free the gastrocnemius and tibio-fibula bone from knee joint to sesamoid bone.

Cut through the Achilles tendon insertion of the gastrocnemius at the sesamoid bone (*cut 4 Fig. I.9*). Continue, by cutting through the tibio-fibula as near as possible to the knee joint (*cut 5 Fig. I.9*) also severing the connection at the knee joint of the anterior tibial muscles. Cut the femur as near to the knee joint as you can (*cut 6 Fig. I.9*).

Carefully remove the preparation consisting of the gastrocnemius muscle attached to the knee joint and sciatic nerve with pieces of vertebrae and, place it in the Petri dish filled with Ringer.


*Materials required for each demonstration*
> Muscle-nerve preparation (see above)
> Kymograph with drum covered with smoked paper
> Muscle mount or chamber
> Piece of cotton
> Thin wire e.g. fuse wire
> Lever assembly

Stimulator—induction coil or electronic stimu-
lator
Key
Dry cell or 2 V accumulator
Cotton covered wire sufficient to complete
circuit
Stop watch or signal marker tuning fork
Frog Ringer
Pipette or syringe
1 stand (two if the tuning fork is used) and
clamps
Pins

*Procedure*

It will depend on the apparatus you are using as to how closely you
follow the description of the arrangement given below. However,
the general grouping of the apparatus for this experiment shown in

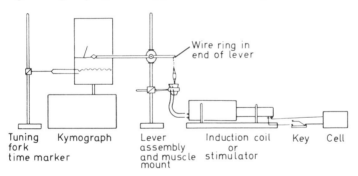

Fig. I.10. *Arrangement of apparatus for the nerve-muscle preparation experiment*

the diagram (*see Fig. I.10*) will apply to all types of apparatus
available, so will the exercises suggested.

If you are using the muscle mount and lever described above,
before arranging the nerve-muscle preparation, tie one end of the
cotton to the Achilles tendon. Arrange the muscle and nerve on the
mount (remember to keep both moist with frog Ringer) and tie the
other end of the cotton to a small wire ring that has been inserted
through the end of the lever (*see Fig. I.10*). The cotton should be long
enough to hold the muscle in a vertical position and allow the lever
to remain horizontal. The lever may need to be weighted. Complete
the electric circuit with the key open. If a metronome stimulator or

make and break contact on the kymograph is used have these ready to include in the circuit. At intervals apply Ringer to both nerve and muscle.

When the tracing nib has been applied to the smoked paper, switch on the kymograph so that a straight line is traced on the drum, shown on the diagram as *A* (*see Fig. I.11*). With the drum at rest complete

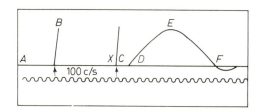

Fig. I.11. *Smoked paper showing trace of muscle twitch*

the circuit with the key. (If a kymograph contact is used it may be necessary to turn the drum manually until it is operating.) By altering the position of the secondary coil of the induction coil with respect to the primary coil adjust the strength of the stimulus. If too weak a stimulus is given it will fail to affect any of the nerve fibres and the muscle will not contract. As the stimulus strength gradually increases and the key is used at each adjustment a muscle contraction will occur that will increase to a maximum (*see B in Fig. I.11*). Further increases in the stimulus strength at the break position will then have no effect. Leave the coils in this position for the duration of the exercise. New smoked paper will be needed for each exercise.

(1) Start the kymograph with the drum revolving at a fast speed e.g. 640 mm/s and stimulate the nerve. The muscle will contract and produce a typical twitch pattern which will be recorded on the smoked paper (*see Fig. I.11*). The pattern shows a latent period (*CD*), a contraction period (*DE*), and a relaxation phase (*EF*). The arrow indicates the point of stimulation.

Unless it is possible to record the exact point of stimulation the duration of the latent period may be difficult to determine. If a kymograph contact is used, the point of stimulation can be recorded before the experiment is started. With the circuit completed rotate the drum by hand stopping at the point when the make or break contact on the kymograph is seen or heard to operate. At this point the muscle will contract and a trace will be produced (*see B Fig. I.11*). It will be at this point (*X*) each time the drum revolves that a stimulus will be effected. The drum motor can then be switched on and the experiment started.

(2) After applying a stimulus and the muscle has completed a contraction apply another stimulus approximately 3/10 s after the first stimulus. What reaction is observed?

(3) At the height of one contraction apply a second stimulus. Try to account for your observations.

(4) Apply two stimuli approximately 1/50 s apart. Account for the curve you obtain.

(5) With a slower drum speed, e.g. 25 mm/s, investigate the effect of repeated stimuli at increasing rates of application e.g. 2/s, 3/s, 8/s and higher rates, until the responses are fused into a sustained contraction known as a tetanus.

*Further work*

Investigate the effect of temperature on the twitch by applying frog Ringer at 3°C, 20°C and 30°C to both muscle and nerve.

# PART J

# Excretion

Excretion in animals is a difficult process to investigate in a school laboratory. The structure of excretory organs may be determined by dissection and microscopical examination. There is considerable variation in the structure of excretory organs, but they all have the function of collecting and removing from the animal the waste products of metabolism. The chief excretory organs of vertebrates, owing to their relatively large size, are usually seen during a dissection of the abdomen. Those of invertebrates are less noticeable but the nephridia of earthworms and the Malpighian tubules of the larger insects, e.g. the locust, are clearly visible and should be examined.

The main nitrogenous waste products of protein metabolism are ammonia, urea and uric acid.

Most aquatic animals including many fishes excrete ammonia as the nitrogenous end product. The toxic nature of this substance demands that it is rapidly removed from the body, an aquatic animal has no difficulty in doing this as there are readily accessible supplies of water into which ammonia may be excreted.

Amphibia excrete mainly urea and so avoid the danger of being poisoned by ammonia when they are living on land.

Some terrestrial reptiles (snakes and lizards) and birds excrete mainly uric acid, a very insoluble product which can be disposed of without the loss of too much water. This is a great advantage to land animals where the problem is how to conserve as much water as possible.

In mammals the chief nitrogenous end product of protein metabolism is urea.

There are many other types of excretory product originating as by-products of all the reactions which constitute metabolism. For example, the reactions involved in tissue respiration result in the production of carbon dioxide and water and when these substances are formed in excess of body requirements, they are excreted to maintain an internal environment of constant composition. Excretory organs are thus important organs in the homeostatic process.

The excretory products of some animals may be collected, human urine is readily obtainable, and analyses may be performed to determine the chemical nature of the constituents. Quantitative estimates of excretory products e.g. urea could be related to the diet e.g. the weight of protein consumed.

## 1. INSTRUCTIONS FOR THE REMOVAL AND EXAMINATION OF A NEPHRIDIUM FROM THE EARTHWORM *(LUMBRICUS TERRESTRIS)*

The tubular excretory organs of *Lumbricus terrestris* are the nephridia. These are paired and are found in every segment except the first three segments and the last one.

*Materials required—Lumbicus terrestris*
$0.65\%$ saline
Dissecting dish
Dissecting instruments
Pins
Killing jar with chloroform pad
Lens (a mounted dissecting lens would be useful)
(Optional—$50\%$, $70\%$, $90\%$ and 'absolute alcohol', borax carmine, xylol, Micrex)

*Procedure*

(1) Kill an earthworm by exposing it to chloroform vapour.
(2) With the worm, dorsal side uppermost, in a dissecting dish place a pin through the anterior end. Slightly stretch the worm and place a pin through the posterior end.
(3) Use fine pointed scissors to open the worm just behind the clitellum. Continue the incision to the anterior and posterior ends

of the body. Pin out the lateral flaps of the body wall exerting slight tension on the septa.

(4) Pipette 0·65% saline over the internal organs to prevent them drying.

(5) You may need to use a lens to identify the coiled tubes of the nephridia (*see Fig. J1*).

(6) Remove one nephridium and the septum to which it is attached as follows. You will probably need to do this dissection under a mounted lens.

With fine scissors cut through the muscular tube close to the body wall.

With fine forceps hold the septum and with scissors in the other hand make a vertical cut through the septum to separate it from the

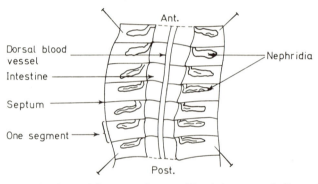

Fig. J1. *Diagram of part of the intestinal region of* Lumbricus terrestris *dissected from the dorsal surface showing position of the nephridia*

gut wall. Then with the blades of the fine scissors held in a horizontal position cut through the part of the septum attached to the body wall.

(7) Place the nephridium and septum on a slide with enough saline to prevent drying.

(8) Examine under low power and use a text book diagram to help you identify the nephrostome and the four different regions of the tube.

(9) Examine under high power and you should be able to see cilia beating in the narrow and the brown regions of the tube. Note the blood capillaries surrounding parts of the tube. Make drawings to record what you see.

(10) If your mounted nephridium clearly shows all the main regions make a permanent stained preparation of it by fixing it in 70% alcohol and staining in borax carmine.

## 2. EXAMINATION OF THE MALPIGHIAN TUBULES OF *LOCUSTA*

*Materials required—Locusta*
Killing vessel
Chloroform
Ether
Wax bottomed dissecting dish
Dissecting instruments
Pins
Dissecting lens
Microscope
Microscope slide

*Procedure*

(1) Kill the locust by exposing it for about 15 min to chloroform and ether vapour in a killing vessel.

(2) Place the locust dorsal side uppermost in a wax bottomed dissecting dish. Anchor by placing pins through the legs and parts of the wings if necessary. Slightly stretch the abdomen and put a pin through the external genitalia.

(3) With fine scissors cut along the mid dorsal line of the abdomen from the anterior to the posterior end. Pin out the cut terga on each side.

(4) Cover with water.

(5) Expose the alimentary canal and identify the various regions.

(6) At the junction of the ileum with the mid gut is a constriction, the pyloric valve. At about this point arise groups of very fine tubules, the Malpighian tubules. Use a lens to see these clearly.

(7) Note the anterior and posterior extent of the tubules and make a drawing to show their position in the abdomen.

(8) Remove one or more of the tubules and examine under a microscope. You should be able to see that each tubule is provided with a small trachea which is wound round it.

## 3. TEST FOR UREA

An enzyme urease present in the seed of the jack bean will convert urea to ammonia and carbon dioxide. Urease is specific in that it will not affect any other nitrogenous product. Use is made of this fact in detecting the presence of urea.

This reaction may also be used as a basis for a quantitative estimation of urea in urine (*see Part J4*).

*Materials required*—4 test tubes in a rack
Filter funnel and filter papers
2 conical flasks
Distilled water
Dilute hydrochloric acid
Dilute sodium hydroxide solution
Phenol red solution (pH 6·8 yellow—pH 8·4 red)
Urea
Urease—The meal containing urease or urease tablets can be obtained from Griffin and George Ltd.

*Procedure*

If jack bean meal is used as the source of urease, some hours before the experiment add 10 cm³ of distilled water to 1 g of jack bean meal in a conical flask. Agitate thoroughly from time to time and then filter. The filtrate is a urease extract.

In a test tube dissolve 0·2 g of urea in 5 cm³ of distilled water. Add five drops of phenol red and one drop of dilute hydrochloric acid to make the solution slightly acid. It must then be made neutral as accurately as possible so that phenol red will indicate the change to an alkaline solution as the urease affects the urea. This is done by pouring half of the solution into a second test tube. To the contents of this second tube add dilute sodium hydroxide solution drop by drop so that the colour becomes just red. Carefully add some of the contents of the first tube to that of the second so that the red colour disappears. Discard the remaining contents of the first tube. To the solution in the second tube add 3 cm³ of the prepared urease extract and five drops of phenol red. Allow the test tube to stand in a warm room. A change in the colour of the solution to red shows that an alkaline solution has developed as a result of the action of urease on urea.

A control tube should be set up using a boiled extract of urease.

## 4.  ESTIMATION OF UREA IN URINE

*Materials required*—0·3% Potassium dihydrogen phosphate solution (dissolve 0·075 g in 25 cm³ distilled water)

Jack bean meal (obtainable from Griffin & George and B.D.H. Ltd.)

N/100 Hydrochloric acid

N/15 Hydrochloric acid

3 thick walled test tubes with stoppers

Three 25 cm³ conical flasks

Burette and stand

Small filter funnel and filter paper

Beaker

Urine

Bromo-cresol green (pH 3·6—5·2 yellow to red)

Methyl red (pH 4·2—6·3 red to yellow)

Distilled water

*Procedure*

Weigh out 1 g of jack bean meal. Mix this with 20 cm³ of the 0·3% $KH_2PO_4$ solution in a conical flask. Shake from time to time for half an hour. Filter the mixture into a second conical flask. Take 3 cm³ of this urease extract in the third conical flask and add three drops of bromo-cresol green and four drops of methyl red. Now very carefully, drop by drop, add N/100 hydrochloric acid until a red tinge is seen. At this point after shaking the solution should be mauve. If too much acid has been introduced add some more of the urease extract.

Dilute 1 cm³ of urine to 10 cm³ with distilled water. Put 2 cm³ of this solution into each test tube. Add three drops of bromo-cresol green and four drops of methyl red. Add N/100 hydrochloric acid until the first tinge of red is seen. Now add to each tube 0·5 cm³ of the neutralised urease solution. Stopper each tube and place in a water bath maintained at 45°C for 7 min. Cool under a cold tap. Take out the stoppers and titrate the contents of each tube with N/15 hydrochloric to produce a first tinge of red. Shake the tubes well during the titrations. Record the volume of hydrochloric acid used in the three titrations. 1 cm³ of N/15 HCl is equivalent to 2 mg of urea.

Calculate the urea content of 100 cm³ of urine.

Calculate the amount of urea excreted each day if it is assumed the volume of urine produced in 24 hours is 1500 cm³.

Would the amount of urea vary as the amount of protein ingested increased or decreased? If you have time compare the amount of urea present in urine specimens from subjects who have had (*a*) large, (*b*) small protein intakes over the previous two days.


## 5. TESTS FOR SOME INORGANIC CONSTITUENTS OF URINE

*Materials required*—Urine
Concentrated hydrochloric acid
Concentrated nitric acid
Ammonium molybdate solution
Silver nitrate solution
5% Barium chloride solution
Bunsen burner
Test tube holder
Test tube rack with test tubes


(1)  *Test for the presence of sulphates*

Add 2 cm³ of concentrated hydrochloric acid to 10 cm³ of urine then add 4 cm³ of barium chloride solution. Should a sulphate be present a precipate of barium sulphate is produced varying in density according to the relative amount of sulphate present in urine.


(2)  *Test for phosphates*

To 10 cm³ of urine add 3 cm³ of concentrated nitric acid and then boil the mixture. Carefully add 5 cm³ of ammonium molybdate and boil. A yellow crystalline precipitate of ammonium phosphomolybdate shows that the urine contains phosphates.


(3)  *Test for chlorides*

To 5 cm³ of urine add a few drops of concentrated nitric acid. To this mixture add 5 cm³ of silver nitrate solution. Should chlorides be present a precipitate of silver chloride will be produced.

The nitric acid is necessary to avoid any urates present being precipitated by the silver nitrate.

## 6. MUREXIDE TEST FOR URIC ACID

*Materials required*—Uric acid (or dried bird excreta—the white part
of the dropping only)
Concentrated nitric acid
2% ammonia solution
Concentrated sodium hydroxide solution
Evaporating basin
Tripod
Gauze
Bunsen burner

*Procedure*

To a small amount of uric acid (or bird excreta) in an evaporating
basin, add a few drops of concentrated nitric acid. Evaporate carefully until the mixture is dry. Allow to cool. To the residue add a drop
or two of 2% ammonia. The substance should turn reddish violet.
Add two or three drops of concentrated sodium hydroxide solution.
The colour changes to blue-violet.

CHRIST'S COLLEGE
LIBRARY

# Response to Environmental Stimuli

The environment of an animal is never static, changes, where they can be perceived, take the form of stimuli to which responses may be made. Stimuli from the environment are divisible into five categories: changes in light intensity and wavelength, changes in the concentration of chemical substances, changes in temperature, the effect of gravity and positional changes of some part of the environment.

Stimuli affecting animals may produce movement patterns in response, the movement will normally be beneficial to the animal. Such a movement may be in a definite direction either towards the source of the stimulus or away from it. This type of directed movement is termed a **taxis.** Taxes may be described according to the nature of the stimulus. For example, when a change in light intensity is the stimulus the movement which an animal makes in relation to the source of stimulus is known as a **phototaxis.** If the movement is towards the light source the animal is said to be positively phototactic and if the movement is away from the light the animal is described as negatively phototactic.

If a stimulus produces a non-directional movement relative to the source of stimulus it is an example of a **kinesis.** If the stimulus causes the animal to make changes of direction as it moves then it is termed **klinokinesis.** If the stimulus results in a speeding up or slowing down of the activity of the animal it is an example of **orthokinesis.**

Stimuli from the environment may cause responses other than movement responses, for example, an animal may respond by showing a colour change when there is a change in the colour or pattern of the background of its environment.

# 1. THE EFFECT OF LIGHT ON A FLAGELLATE CULTURE

*Materials required*—One 2·5 cm diameter specimen tube 10 cm deep
Black card
A rich *Euglena* culture (it has been shown that if the culture from the suppliers is placed into a 250 cm³ beaker containing 200 cm³ of distilled water and 5 cm³ of the liquid from an infusion of horse manure and water, it can be maintained and increased. Add 2 cm³ of the infusion each week and keep in a light room)

*Procedure*

A piece of black card should be cut and glued so that it forms a close but loosely fitting sleeve round the specimen tube (*see Fig. K1*). A top should be cut to fit over this cylinder so that when the card

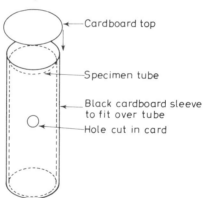

Fig. K1. *Apparatus to demonstrate the effect of light on a flagellate culture*

Cardboard top

Specimen tube

Black cardboard sleeve to fit over tube

Hole cut in card

cylinder and top are in place light is excluded from the specimen tube. A hole 0·5 cm in diameter should be cut in the position indicated in the diagram.

The specimen tube should be almost filled with water containing the *Euglena* culture and stood in a light position (not in intense sunlight). The card cylinder should then be placed over the specimen tube, the hole facing the light, and the card top placed on the cylinder.

The tube should be left for at least a day. Without moving the specimen tube, take off the card top and remove the card cylinder from around the specimen tube.

Note the condition of the *Euglena* culture in the specimen tube. Has the restricted light entry affected the distribution of the *Euglena*?

*Further work*

By arranging a black cardboard cylinder and top for a small crystallising jar as for the specimen tube above, the effect of different coloured light on *Euglena* may be noted. Several holes should be cut in the cardboard cylinder (*see Fig. K2*), and over each, a piece of

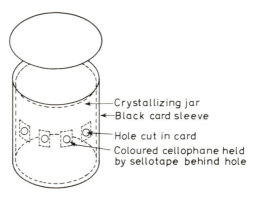

Fig. K2. *Apparatus to investigate the effect of different coloured light on a flagellate culture*

different coloured cellophane (from Griffin & George) should be mounted. To appreciate any gradation in concentration in the *Euglena* clustered near the light sources colours should be in the order Red, Yellow, Green and Blue.

Instead of coloured cellophane try using different thicknesses of tracing paper over each of the holes and noting if different light intensities will affect *Euglena*.

## 2. RESPONSES OF A TURBELLARIAN TO DIFFERENT LIGHT INTENSITIES

*Materials required*—Petri dishes

1 small stone/black enamel paint

Several flatworms, e.g. *Polycelis felina*. Flatworms may be located under leaves and stones in fresh water. (A trap can be arranged in flowing water by placing a piece of raw meat in a small conical flask and submerging this in the stream)

Water from pond or stream

Camel-hair brush

*Procedure*

If a plastic Petri dish is used for the flatworms the base of the dish may be marked with unit squares using a pointed instrument, e.g. a mounted needle and a ruler. Engrave a second Petri dish in a similar way. The dish which contains the flatworms should be stood on a plain white background. Glass Petri dishes should either stand on graph paper or have a piece of graph paper stuck on the underside of the base.

Alternative ways of giving the worms a choice of light intensities are either to place in the water an object, e.g. a stone, with overhangs providing shady positions or to paint an area of the lid of the Petri dish with black enamel paint. If the latter method is adopted the light source should be above the Petri dish.

The Petri dish should contain water from the source from which the worms were obtained.

By means of a clean camel-hair brush transfer the worms to the dish arranging them round the border.

Observe the movements of the flatworms. Record the patterns of movement either by plotting on graph paper or if using an engraved plastic Petri dish use a felt tipped pen to draw the path of movement on the upturned base of a second Petri dish.

*Discussion*

If your planarian showed some response could this be described as a taxis or a kinesis?

Do you consider there are any advantages gained by the animal in its natural habitat by this type of response?

## 3. REACTION OF *LUMBRICUS* TO DIFFERENT LIGHT INTENSITIES

*Materials required*—1 glass T-tube (suggested size: each arm length 5 cm, external diameter 1 cm. Obtainable from Philip Harris)

Black paper

Earthworm (the size chosen to be such that it can easily move in the glass T-tube used)

2 rubber bungs to fit open ends of arms of T tube

*Procedure*

One arm of the T-tube should be covered with a tightly fitting sleeve of black paper. A bung should be inserted in the open end of each side-arm (*see Fig. K3*).

Place the T-tube in a light place with the covered arm on the left and the uncovered arm of the same cross piece on the right. Introduce

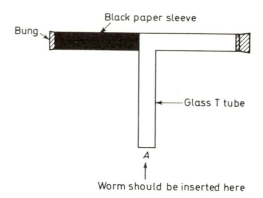

Fig. K3. *Arrangement of T-tube to demonstrate the reaction of* Lumbricus *to different light intensities*

the anterior end of the worm into the open end of the part of the T-tube labelled *A* (*see Fig. K3*).

Note into which of the arms of the cross piece of the T-tube the worm eventually moves.

What type of phototactic response does this choice indicate?

*Further work*

If time is available, repeat the introduction of the same earthworm into the T-tube twenty different times at least, with the tube remaining in the same position as in the above experiment. After this, turn the T-tube so that the darkened side is on the right and the uncovered side on the left. Introduce the worm into the reorientated tube. If the choice of side arm is now different from that for the experiment above how can you explain it?

## 4. PHOTOTACTIC RESPONSES OF *PORCELLIO SCABER*

*Materials required*—Several woodlice, *Porcellio scaber*

These may be found in gardens under large stones and loose bark. If obtained some time before the experiment they can be kept in a plastic aquarium which has a layer of sawdust on the bottom, two or three halves of potato and a piece of bark. Cover the aquarium with glass allowing an access for air exchange

2 plastic Petri dishes

Chloroform

Black enamel paint

*Preparation of choice chamber*

The choice chamber can easily be prepared from two plastic Petri dishes. About 3 cm of the wall of the lower halves of the dishes should be removed. The plastic is usually quite brittle and the operation should be done carefully. The plastic can be scored with a sharp instrument and broken with the aid of pliers (*see Fig. K4*).

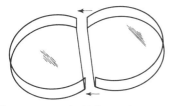

Two prepared dishes should be joined in this way to make lower half of the choice chamber

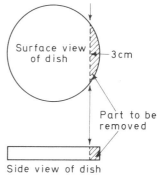

Surface view of dish — 3 cm

Part to be removed

Side view of dish

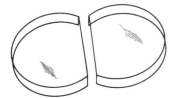

Lids made in similar way to chamber base halves but not joined

Fig. K4. *Stages in the preparation of a choice chamber*

The two open parts should be held coincident and the edges joined by touching each edge of the plastic with chloroform and quickly pushing them together and holding for a minute or so in place. It has been found that if a knife blade or hacksaw blade is heated and this is allowed to touch the cut edges of each half of the Petri dishes and the two are rapidly pressed together the join will be a good one. This operation will allow a passage between two halves of the chamber. The tops should be cut so that when in position over the paired lower halves they fit neatly (*see Fig. K4*). The top half of one side should be painted with black enamel, the side may be painted black but a black paper extension should be added to this side so that it completely covers the lower half of the Petri dish.

*Procedure*

The chamber should be placed in position in an even light. The top half of the chamber which has not been painted should be removed and three or four woodlice placed in the lower half, the top then being replaced.

If the woodlice have been kept in dim light they will quickly respond to the brighter light and explore the available space.

Note if they move towards the half of the chamber which is in darkness. Woodlice are generally negatively phototactic and also exhibit thigmokinesis and if they come together as a group, activity may cease.

*Discussion*

What form of response is it, if when the woodlice are brought from dim to bright light, their movement in the light half of the chamber is rapid but non-directional?

What value could the grouping together be even in light conditions?

Suggest reasons why woodlice are negatively phototactic.

## 5. EXAMINATION OF THE CHANGES IN THE CHROMATOPHORES OF *XENOPUS* WITH A CHANGE IN THE BACKGROUND

A chromatophore is a cell carrying pigment. Those carrying the black pigment melanin are called melanophores and it is these which are to be examined.

Melanophores may be seen in the stretched web of the foot of *Xenopus*. Melanin may be concentrated in the melanophores so that it covers a relatively small surface area, or the pigment may be more widely dispersed thus darkening a greater surface area.

The most convenient way to record the appearance of the melanophores is to use a melanophore index (*see Fig. K5*).

Fig. K5. *Diagram to show the difference in the distribution of pigment within melanophores*

In this example the lowest number represents a cell with melanin concentrated in a small space, the highest number representing the cell where the pigment is more dispersed.

*Materials required per pair of students—*
> 1 *Xenopus* which has been in a black polythene tank for 24 h
> 1 *Xenopus* which has been in a white polythene tank for 24 h
> (The water in each tank should be maintained at the same temperature)
> Tanks obtainable from:—
> Philip Harris Ltd.,
> 63, Ludgate Hill, Birmingham, 3
> Binocular microscope
> 1 net to catch *Xenopus*
> Clean, rough towelling/cotton surgical gloves to hold *Xenopus*

*Procedure*

Two students will be needed to hold *Xenopus* in position for the examination of the melanophores.

Care must be taken to see that the toads do not escape when being lifted from the tank. The easiest method of capture is to use a net to catch *Xenopus*. Before lifting the net clear of the tank cover the opening of the net with a gloved hand (clean cotton surgical gloves are suitable) or with a piece of clean rough towelling. Hold the toad firmly either in a gloved hand or in the towelling just behind the fore limbs leaving one hind leg free. The second student will be needed to help spread out the membrane between the toes on the microscope stage.

Observe the state of the melanophores of each toad using the index to record your observations.

Put the toad from the black tank into the white tank and the one from the white tank into the black tank.

Observe the melanophores every 20 min, replacing the toads in the tanks between observations.

Continue your observations for 2 h or until there is no further change in the appearance of the melanophores.

*Discussion*

If a change in the colour of the background produced a change in the distribution of melanin suggest possible pathways by which this response was made.

How would you test your hypotheses?

## 6. THE AVOIDING REACTION OF *PARAMECIUM* IN RESPONSE TO THE PRESCENCE OF CHEMICALS

The avoiding reaction of *Paramecium* is a response given as a result of stimulation by unfavourable environmental factors. The animal stops moving forwards, moves backwards for a short distance and then moves forwards in a slightly different direction. This pattern of action is due to a momentary cessation of ciliary beat and this is followed by a reversal in beat. The strongly beating cilia of the oral groove are responsible for turning the anterior end of the animal towards a new path. This reaction may be repeated several times before the animal is clear of the unfavourable stimulus.

The avoiding reaction has survival value as it enables *Paramecium* to move away from unfavourable areas and remain in areas favourable for its existence.

*Materials required*—A rich culture of *Paramecium*
Microscope (a binocular is useful as the large field of vision enables the patterns of movement of many individuals in the culture to be seen at one time)
Cavity slides
Fine bore pipettes with teats

*Suggested chemicals—*

> 0·05% acetic acid
> 0·05% hydrochloric acid
> 0·5% sodium chloride

*Procedure*

Pipette a few drops of a rich *Paramecium* culture on to a cavity slide. Examine with a microscope and note the types of movement taking place.

With a fine pipette place one droplet of one of the above solutions at one edge of the culture on the slide. Watch the behaviour of *Paramecium* on adding the test solution and note any changes which take place in their distribution after several minutes.

Make a simple diagram to show the distribution of *Paramecium*.

Repeat this procedure for the other solutions using clean slides and pipettes and fresh drops of culture each time.

After watching a single individual showing the avoiding reaction, make diagrams to illustrate this response.

*Further work*

*Paramecium* is said to collect in weakly acidic regions. By making up more dilute solutions from your stock 0·05% hydrochloric acid solution try to find out which concentration appears favourable.

*Response to other stimuli*

Devise experiments to investigate the response of *Paramecium* to mechanical stimuli resulting from contact with objects placed in the drop of culture.

The response to regions of different temperatures of the culture medium could also be investigated.

## 7. CHEMOTACTIC RESPONSES OF A TURBELLARIAN

*Materials required*—Several flatworms—These should have been isolated and starved for six days. The collection should be examined several times a day during this period and any dead specimens immediately removed (*see Part K2* for collecting suggestions)

> Several living *Gammarus* or a 1 cm³ piece of raw meat or liver. Fresh water shrimps can be collected at the same time as the flatworms
> 2 Petri dishes
> Glass rod
> Camel-hair brush

*Procedure*

If using *Gammarus*, place one in each of the two bottom halves of the Petri dishes which have been half filled with water. Using a glass rod crush the shrimp in one of the Petri dishes. Immediately, using the camel-hair brush, transfer one flatworm to each dish. The dish containing the living shrimp and flatworm will act as the control.

If using meat, place it in the centre of a Petri dish containing some water. Transfer a single flatworm into this Petri dish near the border.

Note carefully the response of the flatworm in the presence of the foodstuffs. What does the flatworm do if and when it reaches the food?

Movement of the flatworm may be recorded as in *Part K2*. The mucous track made by the animal on the base of the Petri dish may be 'developed' by sprinkling a little talcum powder on the base at the end of the experiment when the contents of the dish have been removed. The powder should be washed with a little water by rocking the Petri dish. The tracks now made visible can be checked against details recorded on the graph paper.

## 8.  DEMONSTRATION OF KINESES IN *PORCELLIO SCABER*

*Materials required*—1 choice chamber—This should be prepared as described in the experiment K4. The bases of each half of the chamber should be punctured a number of times with a hot needle. Each hole should be about 1 mm in diameter. (It is suggested that a square of holes, 12 rows each of 12 holes with 2 mm between each hole, is produced.) Neither of the two halves comprising the top should be painted.

> 2 glass Petri dishes each about two-thirds the diameter of one half of the bottom of the choice chamber
> Concentrated $H_2SO_4$
> Woodlice, *Porcellio scaber*

*Procedure*

One of the smaller Petri dishes should be carefully half-filled with concentrated $H_2SO_4$. The other small dish should be half-filled with water.

The choice chamber should be arranged on the two smaller Petri dishes, one half over that containing $H_2SO_4$ and the other half over the dish containing water (*see Fig. K6*).

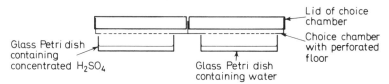

Fig. K6. *Arrangement of choice chamber to investigate kineses*

5 woodlice should be placed in each half of the chamber which can then be covered.

Notice the general pattern of response movement in the woodlice and their eventual resting place.

If it is wished to follow any particular animal the dorsal surface may be marked with a spot of coloured cellulose paint to facilitate rapid identification at any time.

*Further work*

(1) Instead of covering the chamber with the ordinary covers use black painted ones. After inserting the woodlice and covering, remove the covers after 5 min and notice the position of the woodlice.
(2) Substitute water at different temperature extremes (0°C and 20°C) in each of the small dishes and repeat the experiment noting carefully the relative activity of the woodlice in each half of the choice chamber and their eventual resting place.

# Growth. Development. Regeneration

## GROWTH AND DEVELOPMENT

When an organism grows it undergoes a relatively permanent change in size or form. The changes will usually be in a positive direction resulting in an increase in the size of the organism over a period of time.

The rate of growth of an organism is normally found by making regular measurements in one dimension e.g. length or circumference or by regularly weighing the organism. This latter method has obvious drawbacks as temporary changes in water content or storage products may obscure the more permanent changes in weight. A measurement of the dry weight of an organism is preferable but obviously this is not possible for individual organisms. Dry weight measurements may be used quite satisfactorily for estimations of the average growth rate of a particular population. Samples being withdrawn at regular intervals and dry weight measurements made.

All the parts of an animal do not grow at the same rate. Thus with a change in the size of the animal as a whole there will be a corresponding change in the proportion of various parts of the anatomy.

If the growth rate of some part of an animal is the same as the mean growth rate of the whole animal, growth of that particular part is said to be isometric. Growth is allometric if the growth rate of the specified part differs from the mean growth rate of the whole animal.

During the life history of an animal the growth rate will vary during the different phases of the life span. If growth is plotted against time a growth curve is obtained. In many animals the growth curve is sigmoid (*see Fig. L1*).

That is, the growth rate is slow at the start of life, it then becomes more rapid and when the animal is adult remains at a zero level until the period of senescence is reached when growth becomes negative. Each species has a characteristic growth pattern, those animals which continue to grow throughout their adult life will show a

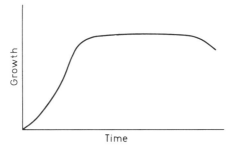

Fig. L1. *The sigmoid growth curve*

growth curve differing from the sigmoid curve described above. In this group are many of the invertebrates with the exception of the insects, and some vertebrates including many fishes. The growth rate of these animals decreases with age but never quite reaches a zero value (*see Fig. L2*).

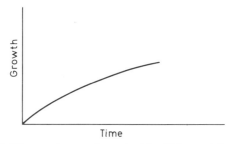

Fig. L2. *The growth curve of an animal in which growth is unlimited*

The rate of growth may be affected by different external factors, for example, the amount of nutrients available or the environmental temperature, and by internal factors e.g. the growth stimulating hormones produced by the thyroid gland and the pituitary gland.

Growth in the young animal is accompanied by development. When an organism is said to be in a state of development, it is implied that structural and physiological changes are taking place to gradually convert the young animal to the mature form.

In some animals, during the period of growth and development to the adult, a larva is formed. The term larva usually describes a young self-supporting form whose appearance and habits differ from those of the adult. The stages involved in the transition of a larva to an adult are collectively called metamorphosis.

PREPARATION FOR PRACTICAL WORK

Investigations involving growth and development are obviously relatively long term projects. Weeks or even months may be needed for a series of observations or measurements. It is important to bear this in mind so that your investigations are planned to start at a convenient stage in the course.

## 1. INVESTIGATION OF THE RATE OF GROWTH OF *CARAUSIUS MOROSUS*

In this species of stick insect mature females produce eggs parthenogenetically.

The eggs are clearly distinguishable from faeces. Each egg has a dark brown hard shell which is rounded at one end and has a pale coloured cap at the other end. Each egg is just over 2 mm long.

The length of the incubation period varies with environmental conditions. You should begin to collect eggs from the culture about 2–3 months before you need the nymphs. As mature female stick insects lay their eggs over a period of weeks a culture containing several mature forms will enable a number of investigations to be performed at intervals.

When the eggs are collected the date of collection should be recorded.

*Materials required*—Stock of female *Carausius morosus* supplying eggs

Source of fresh privet or ivy leaves. You may find that the leaves of other plants e.g. the blackberry, are equally acceptable

Stock container, either an insect cage or a small covered aquarium

Small holder for plant stems (small paste jars are suitable)

Cotton wool to plug the neck of the food container

*Additional requirements for each student—*

> 1 newly hatched nymph of *Carausius*
> 1 wide mouthed jam jar covered in a way which will allow access of air but prevent escape of the insect
> 1 small container to hold a plant stem and which fits into the jam jar
> 1 pair of dividers
> Ruler

*Procedure*

Use the dividers to make the following measurements weekly:
(1) The length of the body from the anterior tip of the head to the posterior end of the abdomen.
(2) The length of one antenna.
(3) The length of the mesothorax.

Each time record the date on which the measurements were made.

Continue to make measurements for as long as you are able to or until at least 6 measurements give the same results.

*Interpretation*

(1) Record your results graphically, plot length of body against time.

For an example *see Fig. L3*. Account for the shape of your graph.

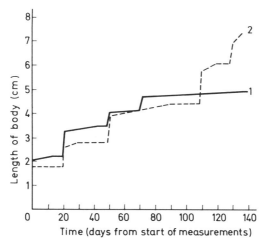

Fig. L3. *Growth curve of two stick insects*

(2) Plot a graph for an antenna and the mesothorax.

(3) Do the antenna and the mesothorax grow at the same rate as the whole body?

(4) Compare your results with other members of the group. As a result of this comparison are you able to reach any conclusions regarding the growth of an average stick insect.

FURTHER WORK TO INVESTIGATE THE EFFECT OF ENVIRONMENTAL FACTORS

There are two developmental periods which may be examined usefully.

(1) *The incubation period*

The effect of different temperatures on the incubation time may be studied.

Does the early incubation temperature influence the body form of the insects which develop?

There are records of 'males' appearing very much more frequently when eggs are incubated at 30°C for the first 30 days of embryonic life. Even short periods (7–14 days) of incubation at 30°C within the first 30 days may produce insects with male characteristics. 'Males' of Carausius morosus are much smaller with relatively longer antennae, they have a narrower thorax and a smooth cuticle.

(2) *The nymphal stages*

Several environmental variables may be investigated, temperature, humidity, amount of food available and light are factors which could be examined.

Where only one environmental factor is being investigated, you should set up both experimental and control cages with environmental conditions differing only in that factor.

## 2. SUGGESTIONS FOR INVESTIGATING THE GROWTH AND DEVELOPMENT OF *LOCUSTA*

*Preparation*

*See Part D3.6* for details of breeding locusts in a laboratory.

*Incubation period*

(1) Investigate the effect of temperature on the length of the incubation period. Incubate the eggs at different temperatures in the range 25°C–35°C.

(2) Is there a change in the weight of the eggs during incubation?

If treated with care individual eggs may be separated from the egg pod and weighed daily during the incubation period. Each day weigh one egg or a sample containing a known number of eggs. After weighing replace the eggs in moist sand.

At the end of the incubation period plot egg weight (mg) against time (days). Try to account for the shape of your graph. Is your result similar to those made for other locust eggs?

*Nymphal period*

(1) Make observations of development after hatching.

In a cage containing nymphs you will probably notice that the growth rate of individual nymphs varies.

Remember that only the adult locust is capable of flight. Note the position and size of the developing wings during the nymphal period.

(2) Rear a small number of locusts in a separate cage. Starting with the first stage hoppers find the total weight of the nymphs each day. For weighing, the nymphs should be placed in a weighed covered container, e.g. a plastic carton. At the end of the nymphal period plot weight (g) against time. Try to account for the shape of your curve.

(3) Investigate the effect of temperature on the length of time spent in each nymphal stage. Note carefully the colouration of the exoskeleton at different temperatures.

(4) Investigate the effect of isolation on the colour of the exoskeleton of the nymph.

One first stage hopper should be isolated and reared separately from the others which hatch from one egg pod. Each day fresh grass should be given and this should be placed in a small jar of water in each cage. The temperature of the two cages should be identical. The isolated hopper should not have any visual contact with the other hoppers. At all stages of development compare the colours of the exoskeleton.

Find out what happens if nymphs at other stages of their development e.g. third stage nymphs are isolated from the crowd and reared separately for the rest of the nymphal period.

What happens if a nymph (e.g. third stage) which has previously been reared in isolation is transferred to a cage containing many other nymphs of the same age?

Remember that your results have been obtained in artificial conditions. If these results are also obtainable under natural conditions suggest what their adaptive significance might be.

## 3.  A SIMPLE METHOD FOR THE EXAMINATION OF SOME OF THE LARVAL STAGES OF *CYCLOPS*

*Cyclops* as a genus is widely distributed in fresh water and may be found in most ponds, lakes and ditches.

There are several features of this animal which make it especially attractive for study. The egg carrying female of the genus is recognisable without a microscope and thus she can be quickly selected from samples of pond water. (*See Fig. L4* for details of the appearance of the female *Cyclops*.)

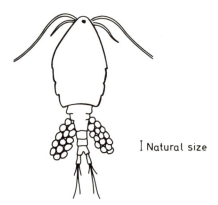

I Natural size

Fig. L4. *Outline diagram of female* Cyclops *(Dorsal view)*

The larvae which hatch from the eggs are known as nauplii. The oval shaped nauplius is provided with three pairs of jointed appendages (*see Fig. L5*). The nauplii, although capable of rapid darting movements, may remain motionless in the water giving sufficient time for microscopic observations of their structure.

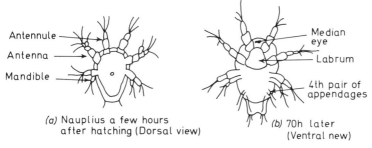

Fig. L5. Cyclops *nauplii (not drawn to scale)*

The nauplii pass through a series of moults, after each moult the larva increases in size and develops more of the adult characteristics.

The time taken for development probably varies with the temperature of the water. You could investigate the effect of this environmental factor on development.

*Materials required*—Adult female *Cyclops* with egg sacs
Pond water from which *Cyclops* was obtained
Cavity slide (cavity 2·1 cm) or watch glass and cover
Microscope
Teat pipette

*Procedure*

Place a female *Cyclops* with egg sacs in pond water in a cavity slide. The cavity slide need not be covered but to prevent dehydration of *Cyclops* the slide should be examined at intervals during the day and more pond water added if necessary to prevent drying. If regular attention is not possible a covered watch glass containing a greater volume of water should be used.

With low power magnification examine the contents of the pond water. Remove all other crustaceans.

Examine the egg sacs. Are you able to see any internal detail of the eggs? Record the date and any observations you make on that date.

After this initial examination continue to observe the contents of the water at regular intervals, at least once every day.

Record the date when you first see the nauplii. What happens to the two groups of egg cases?

24 hours after the appearance of the nauplii examine the water for evidence of the first moult. You may be able to see the cast cuticles and slightly larger nauplii present.

Continue to make observations recording the dates when you notice any change in size or shape of the larvae. Note particularly the number and arrangement of appendages and the gradual change in the shape of the body.

Make drawings to illustrate development.

## 4. DEVELOPMENT OF THE TROUT

The early growth and development of a fish—the trout, may be observed under laboratory conditions. This necessitates obtaining eggs and the construction and maintenance of a simple hatchery.

*Source of eggs*: During January to March trout eggs may be purchased from:—The Midland Fishery,
Nailsworth,
Stroud,
Gloucester.

CONSTRUCTION AND MAINTENANCE OF HATCHERY

As soon as the eggs are received (they are usually packed in damp moss in a tin), they should be placed in an enamel tray such as a dissecting dish. The dish should be arranged as in the diagram (*see Fig. L6*), placed permanently over the edge of a sink and raised at an angle of 15° to the sink. Some thoroughly washed gravel can be used to cover the bottom of the dish.

Water must be kept slowly running into the dish and allowed to overflow at the opposite end. Moving water will help aeration and keep the temperature low (no higher than 10°C), but it must not produce persistent bubbles which are apt to get caught under the gill covers. A device for evenly distributing a flow of water on to the surface can be made from a glass T-tube. On each side of the cross piece a piece of plastic tubing is fixed blocked by a cork or bung at each end. The plastic tubing should be punctured with an awl point to produce a series of holes on each side of the T-tube cross piece (*see Fig. L6*).

At the sink end of the dish a small portion of the top should be carefully covered with net (or plastic or zinc gauze of mesh no greater

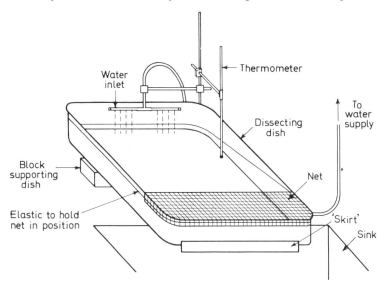

Fig. L6. *Simple hatchery for trout*

than 0·25 cm). This should be cut to shape and held in place with elastic under the rim of the dissecting dish. The net is needed to prevent the alevins from being carried over the end of the dish in the overflow water. Should the overflow water creep underneath the tray and on to the bench a 'skirt' of cloth may be fixed at the end as shown in the diagram.

The incoming water should have a pH value between 6·6 and 7·6. Philip Harris offers an aquarium indicator for this range.

A. S. Gillespie (1953, *Sch.Sci.Review 125*, 96) suggests that the presence of normal amounts of chlorine in tap water does not appear to have too great an effect on the alevins but a dechlorination method can be employed. On no account should the water be delivered through copper piping. A crash in the population of trout alevins may occur if consideration of any of the above factors is neglected. It may be advisable therefore, to have a number of dishes and constantly examine these for any signs that all is not well. Always immediately remove any fish that are dead or eggs that have failed to hatch after the main batch. The viable eggs are 'eyed' showing their black eyes, any that are seen to be partly white and opaque should be removed. The newly hatched trout is termed an alevin (*see Fig. L7*).

The young alevins will not need feeding for from three to five weeks as the yolk present in the yolk sac will suffice. As the yolk sac

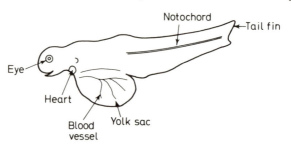

Fig. L7. *Newly emerged alevin*

is seen to get smaller the alevins become more active and can be offered food. Live or dried *Daphnia* can be introduced into the water at this stage. Minced liver can be given but if any remains after feeding it should be removed to prevent the water becoming polluted. It is worthwhile maintaining a *Daphnia* culture (*see Part D3.4*). Free feeding trout having lost their yolk sac are termed fry and at this stage can be moved to a deep tank with adequate aeration. A pump with diffuser to agitate and aerate the water can be used (see above for warning about bubbles). A water filtration and circulating device will improve the chances of survival. A constant supply of zooplankton should be available for feeding the fry.

PROCEDURE FOR EXAMINING THE ALEVINS

During the time the alevins are developing they can be removed at regular intervals and measured. A growth curve can be constructed. Measuring can be done by carefully using a blunt ended pipette (*see Fig. L8*) and placing the young fish in a plastic Peri dish containing

Fig. L8. *Blunt ended pipette*

water from the same source as that used for the hatchery. For measuring convenience the Petri dish may be scored on the under-surface to produce a cm/mm grid system (*see Fig. L9*). The scoring with the point of a sharp awl, or the top of a scalpel blade using a metal ruler as a guide.

The alevin may be anaesthetised to make measuring and observation easier, it should be put into a solution of 1 : 3000 MS.222 for up to 15 min and then transferred to the Petri dish. A dissecting lens or binocular microscope can be used when inspecting the alevin.

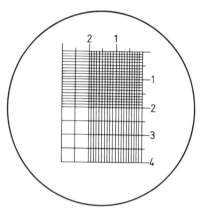

Fig. L9. *Plastic Petri dish the underside of which has been scored to give a cm/mm grid*

Notice also, during the course of the growth of the alevin, the development of the skeletal system. The alevin may be used also to demonstrate the action of the two chambered heart and capillary circulation (*see Part H1.1*).

# 5. LARVAL DEVELOPMENT IN *RANA* OR *XENOPUS*

Development from spawn through the tadpole stages to metamorphosis with the development of an adult form may be observed in both species (*see Part D3.9* for notes on the care of *Xenopus*).

The average growth rate of tadpoles may be found by measuring at regular intervals a sample of 10 tadpoles and calculating the average length. The total length and the length of some specified part being recorded. A comparison of the two rates of growth may then be made.

For measurements of length you may need to anaesthetise the tadpoles. Use M.S.222 (*see Part H1.1*).

Individual measurements may be made with the anaesthetised tadpole in a water filled plastic Petri dish which has a grid engraved on its base (*see Part L4, Fig. L9*).

Tanks of water maintained at different temperatures (try 15°C–25°C range) could be used to investigate the effect of temperature on the rate of growth. The growth rate is probably influenced by other environmental factors e.g. amount of food available, amount of oxygen for respiration, and amount of space, so that it is important that in the experimental tanks all environmental factors except the one under investigation, are identical.

## 6. DEVELOPMENT OF THE CHICK EMBRYO

Where possible chick embryos should be examined daily throughout the period of incubation.

Eggs should be removed from the incubator after a known length of time. Part of the shell is then carefully removed to expose the embryo (see below). External features in the development of the fore and hind limbs, the feathers, the beak, the eyes and the ears may all be seen quite clearly. Note the development and position of the embryonic membranes.

Microscopic examination of the early embryos should be made in order that greater details of development may be seen.

### 6.1. A METHOD FOR MAKING A WHOLE MOUNT PREPARATION OF AN EARLY CHICK EMBRYO.

*Materials required*—Incubator adjusted to maintain a temperature of 38°C

0·75% Saline

Fertile eggs. Order in excess of the numbers required as some eggs may prove to be infertile

Fertile eggs may be obtained from:—

T. R. Dunn and Son Ltd.,
Broad Oak Farm,
High Legh,
Knutsford,
Cheshire.

*Additional requirements for each student*—

Support to hold the egg during the removal of the embryo. A suitable support may be made by shaping a mould in a block of expanded polystyrene

Mounted needle
Fine pointed scissors
Blunt forceps
Teat pipette
Camel-hair brush
Petri dish (both halves)
One 5 cm diameter watch glass with convex bottom
Four 5 cm diameter watch glasses with flat bottom
Microscope
Microscope slide, cover slip and support for cover slip
Bouin's fluid (*see Part C1.1*). Dilute 2 parts Bouin to 1 part distilled water before use
50%, 70%, 90% and 'absolute' alcohol
Borax carmine
Cedarwood oil or clove oil
Xylene
Canada balsam

*Procedure*

Obtain an egg which has been incubated for about 60 h. Place the egg in the supporting mould contained in a dish so that the egg rests in the same position as it was in the incubator.

Pierce the broad end of the shell with the needle so that air escapes from the air chamber thus reducing the internal pressure.

Insert the point only of the scissors and make a horizontal cut right round the egg so that just less than the upper half of the shell may be removed. The embryo should now be visible.

Pour off as much of the albumen as possible and float the yolk and embryo in a Petri dish of 0·75% saline.

With fine scissors make a circular cut well clear of the embryo so that it is separated from the underlying yolk.

With blunt forceps take hold of the separated extra embryonic tissue and transfer this and the embryo to a watch glass containing 0·75% saline.

Remove the vitelline membrane and move the disc of tissue about in the saline to rid the embryo of any attached yolk.

Float the disc on to the convex side of an upturned watch glass and spread the tissues out carefully. Remove from the saline and place the watch glass (still convex side uppermost) in a Petri dish.

Pipette on Bouin's fluid to fix the preparation. Pipette on enough fixative to keep the edges of the preparation from adhering to the watch glass.

After a few minutes detach the preparation and transfer it to fresh fixative in another watch glass. Leave until the embryo has been in fixative for 1 hour.

Wash twice in 50% alcohol until the yellow colour has disappeared.

Stain in borax carmine for 3–4 h.

Dehydrate using alcohol.

Clear in clove oil or cedarwood oil, for 12 h or over night.

Wash in xylene.

Mount in Canada balsam.

Cover with a supported cover slip.

Label.

The preparation is now ready for examination.

## 7. MEASUREMENTS OF THE GROWTH RATE OF INDIVIDUAL MICE FROM ONE LITTER

*Preparation*

*See Part D3.10* for notes on the care of mice in the laboratory and *Part E4* for notes on the life cycle.

*Materials required*—1 litter of mice containing at least 8 young, ideally with equal numbers of males and females, and aged 21 days. More than 1 litter may be needed if each student or pair of students is to be responsible for 1 mouse

2 mouse cages per litter provided with bedding material and water dispensers

Balanced diet

Suitable containers for holding a mouse during weighing. For example, plastic cartons for Yoghurt or cream provided with snap-on lids which should be perforated to allow access of air. These containers are easy to clean after use

Balance

Rulers

*Procedure*

The mice should be sexed and the males and females isolated in two cages.

If the individual mice cannot be distinguished by the coat colour alone they should be marked in some way e.g. by ear clips or the application of a non-toxic dye to the coats.

The 2 cages should be placed near one another so that the two sets of mice will experience similar environmental conditions.

Provide each cage with the same amount of food weekly. The amount of food available should be in excess of requirements so that the possibility of one mouse not receiving enough food is eliminated. (Four adult mice consume about 170 g per week.)

For each mouse 3 different measurements of growth should be made:—

(1) Weight

(2) Length of body (tip of snout to base of tail)

(3) Length of tail

These three measurements should be made once a week for at least 5 weeks i.e. until the mice are 8 weeks old, or longer if possible. Measurements should be made at the same time of day.

*Interpretation*

At the end of your period of measurement for each mouse, use your 3 sets of readings to draw 3 growth curves.

Plot (1) Weight (*g*) against time (*days*)

    (2) Length of body (*cm*) against time (*days*)

    (3) Length of tail (*cm*) against time (*days*)

Are the three curves of similar shape? Between which two dates did the greatest increase in growth occur?

Plot the logarithm of tail length (*y* axis) against the logarithm of body length (*x* axis). If you obtain a straight line whose gradient is 45° the growth rate of body and tail is similar. If the gradient is less than 45° the rate of growth of the tail is less than that of the body, if the gradient is greater than 45° the rate of growth of the tail is higher than that of the body.

Examine the growth curves for other members of the litter. Is there any variation in their shape?

Can you relate any differences which you observe to the sex of the mice? Calculate the average growth in weight and length of each sex and plot growth curves.

For each sex was the smallest mouse at the start of the investigation the smallest at the end of the period of measurement?

The results of this investigation should enable you to make some tentative statements about the growth rate of mice. You will need to perform further investigations before you can reach any general conclusions. You should be able to devise other experiments to help you do this.


## REGENERATION

Most animals possess to a certain degree, the power to replace or regenerate parts of their bodies that have been lost.

The relative extent of the regeneration which can occur appears less marked as the organisation of the body structures becomes more complex. In many of the lower multicellular invertebrates the power to regenerate is impressive, a complete body being produced from as little as a sixth of an original organism. In adults of the higher vertebrates only small parts appear to be restored. Here it may be difficult to distinguish between the considerable interchange of material at the molecular level as a result of the normal metabolic processes, and the renewal and replacement of parts that may be termed regeneration.

Undoubtedly the power of regeneration has a high adaptive value. Regeneration may be associated with asexual reproduction. In the lower invertebrate groups the process of asexual reproduction as a result of budding or fragmentation produces a rapid increase in numbers which confers considerable survival potential. The use of body parts to produce a new organism may necessitate replacement or regeneration of lost parts.

Observation of regeneration processes and the rate of the process in various areas of the body reveals something of body tissue organisation. The regenerative process of a simple animal may be demonstrated by experiments with planarians.


## 8. REGENERATION IN PLANARIANS

*Materials required*—Razor blade or sharp scalpel blade
Plastic Petri dishes
Felt tipped pen or Glassrite pencil
Ice
Rain water or distilled water
Dissecting or binocular microscope
Planarians—suggested species *Polycelis nigra*,
  *Polycelis felina*, *Dugesia lugubris* (*see Part
  D3.2* for methods of capturing planarians)
Camel-hair brush

*Procedure*

Using the camel-hair brush transfer a planarian to some rain or distilled water contained in a Petri dish. Crush an ice cube and place a few pieces of ice into the Petri dish so that the temperature falls to about 2°C. The lowering of the water temperature will slow down the movement of the planarian and make cutting easier.

Carefully arranging the Petri dish containing the planarian under a dissecting or binocular microscope use the sharp scalpel blade to cut the body of the planarian. Make a swift clean cut and use the pattern suggested in the figures. Repeat the same pattern of cut on several more planarians in the same way. Use the felt tipped pen to note on the Petri dish lid the type of cut, species of flatworm and the date. Cover the Petri dishes and keep in a cool shady place. Inspect the dishes each day and remove any dead pieces of worm. Note carefully what happens in each dish and using a lens make diagrams at all stages.

In deciding what cutting pattern to use consider the following questions:—

(1) If a transverse cut is made (*see Fig. L10*) does each half regenerate at the same rate? If not does this suggest a gradient of metabolic

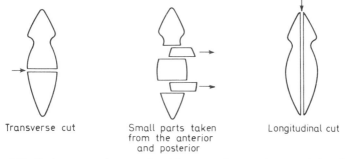

Transverse cut

Small parts taken from the anterior and posterior

Longitudinal cut

Fig. L10. *Cutting patterns in connection with regeneration experiments and planarians*

activity? Does this experiment demonstrate polarity in the tissues?
(2) How large a part of a planarian is necessary to achieve successful regeneration?
(3) What happens if the animal is cut longitudinally into two equal halves?
(4) What happens to small parts that are separated from (*i*) the anterior and (*ii*) the posterior of the animal? (*see Fig. L10*).
(5) If the head is cut down the middle what happens? (*see Fig. L10*).
(6) If you have used more than one species do they react equally and regenerate at the same rate? Which species showed the greatest capacity for regeneration?

244

# APPENDIX 1

FURTHER READING AND VISUAL AIDS

## PARTS A and B

Baldwin, E. (1962). *The Nature of Biochemistry*. Cambridge University Press
Bladwin, E. (1963). *Dynamic Aspects of Biochemistry* (4th edn). Cambridge University Press
Harrison, K. (1959). *A Guide Book to Biochemistry*. Cambridge University Press
Holter, H. (1961). 'How things get into cells', *Scient. Am.* September, 1961. Reprint 96
Kruyt, H. R. and Overbeek, J. T. G. (1960). *An Introduction to Physical Chemistry for Biologists and Medical Students*. London; Heinemann
Ramsay, J. A. (1965). *The Experimental Basis of Modern Biology*. Cambridge University Press
Walsh, E. O'F. (1961). *An Introduction to Biochemistry*. English Universities Press
*The Structure of Protein*. Film on free loan from Unilever

## PART C

Barrass, R. (1964). *The Locust*. London; Butterworths
Peacock, H. A. (1966). *Elementary Microtechnique*. London; Arnold
Thomas, J. (1963). *Dissection of the Locust*. London; H. F. and G. Witherby, Ltd.
Nuffield Biology Text (1967). Year 5. London; Longmans/Penguin Books
Nuffield Film Loop No. 3037. London; Longmans/Penguin Books

## PART D

Bryant, J. J. (1967). *Biology Teaching in Schools Involving Experiments or Demonstrations with Animals or Pupils*. The Association for Science Education
Hunter-Jones, P. (1961). *Rearing and Breeding Locusts in the Laboratory*. London; Anti-Locusts Research Centre
Hunter-Jones, P. (1966). 'Allergy to animals: A zoological hazard', *New Scient.* September, 1966, p. 615
Leadley Brown, A. M. (1970). *The African Clawed Toad*. London; Butterworths
Wilson, R. W. (1968). *Useful Addresses for Science Teachers*. London; Arnold
U.F.A.W. (1967). *Handbook on the Care and Management of Laboratory Animals* (3rd edn). Edinburgh; E. and S. Livingstone, Ltd.
*Humane Killing of Animals*. (1968). Universities Federation for Animal Welfare
*The Ruthless Ones*. (Life History of Locusts). Film on free loan from Petroleum Films Bureau

## PART E

Bailey, N. T. J. (1959). *Statistical Methods in Biology*. English Universities Press
Darlington, C. D. and Bradshaw, A. D. *et al.* (1963). *Teaching Genetics*. Edinburgh; Oliver and Boyd
Mackean, D. G. (1968). *Introduction to Genetics*. London; John Murray
Strickberger, M. W. (1962). *Experiments in Genetics with Drosophila*. New York; John Wiley

PART F

Abbott, D. and Andrews, R. S. (1965). *An Introduction to Chromatography*. London; Longmans
Barrington, E. J. W. (1967). *Invertebrate Structure and Function*. London; Nelson
Borradaile, Eastham, Potts and Saunders (1961). *The Invertebrata* (4th edn revised by G. A. Kerkut). Cambridge University Press
Taylor, R. J. Unilever Educational Booklet, *The Chemistry of Proteins*. Unilever
Films and Film Loops on *Paper and Thin Layer Chromatography*. Shandon Scientific Co. Ltd.

PARTS G, H and J

Best, C. H. and Taylor, N. B. (1954). *The Living Body*. London; Chapman and Hall
Clegg, P. C. and Clegg, A. G. (1962). *Biology of the Mammal*. London; Heinemann
Green, J. H. (1963). *An Introduction to Human Physiology*. Oxford University Press
Ramsay, J. A. (1952). *A Physiological Approach to the Lower Animals*. Cambridge University Press
Rowett, H. G. Q. (1959). *Basic Anatomy and Physiology*. London; John Murray
Yapp, W. B. (1960). *An Introduction to Animal Physiology*. Oxford University Press

PART I

Arthur, D. R. (Ed.) (1966). *Looking at Animals again*. London; W. H. Freeman
Cox, F. E. (1968). 'Parasites of British Earthworms', *Jnl. Biol. Educ.* **2**, 2
Ealing Film Loops: *Amoeba proteus* (81–5019), *Paramecium aurelia* (81–5027/1), *Euglena gracilis* (81–5050/1), *Daphnia* (81–6090)
Hyman, L. (1940). *The Invertebrata*: *Protozoa through Ctenophora*. New York; McGraw-Hill
Vickerman, K. and Cox, F. E. (1967). *The Protozoa*. London; John Murray

PART K

Carthy, J. D. (1966). *The Study of Behaviour*. London; Arnold
Evans, S. M. (1968). *Studies in Invertebrate Behaviour*. London; Heinemann
Thorpe, W. H. (1963). *Learning and Instinct in Animals*. London; Methuen

PART L

Gillespie, A. S. (1953). Hatching Trout in the School Laboratory. *Sch.Sci.Rev.* **125,** 96
Lewis, T. and Taylor, L. R. (1967). *Introduction to Experimental Ecology*. London; Academic Press
Moog, F. (1964). *Animal Growth and Development*, B.S.C.S. Laboratory Block. London; Harrap
Nuffield Biology Text (1967). Year 1. London; Longmans/Penguin Press

PERIODICALS

*Journal of Biological Education*
*New Scientist*
*School Science Review*
*Science Journal*
*Scientific American*

# APPENDIX 2

| Degrees of Freedom | 0·99 | 0·95 | 0·80 | 0·50 | 0·30 | 0·20 | 0·10 | 0·05 |
|---|---|---|---|---|---|---|---|---|
| 1 | 0·00016 | 0·0039 | 0·064 | 0·46 | 1·07 | 1·64 | 2·71 | 3·84 |
| 2 | 0·020 | 0·10 | 0·45 | 1·39 | 2·41 | 3·22 | 4·61 | 5·99 |
| 3 | 0·12 | 0·35 | 1·01 | 2·37 | 3·67 | 4·64 | 6·25 | 7·82 |
| 4 | 0·30 | 0·71 | 1·65 | 3·36 | 4·88 | 5·99 | 7·78 | 9·49 |
| 5 | 0·55 | 1·15 | 2·34 | 4·35 | 6·06 | 7·29 | 9·24 | 11·07 |

Atomic weights of some common elements

| Element | Symbol | Atomic weight | Element | Symbol | Atomic weight |
|---|---|---|---|---|---|
| Aluminium | Al | 26·97 | Manganese | Mn | 54·93 |
| Antimony | Sb | 121·76 | Mercury | Hg | 200·61 |
| Arsenic | As | 74·91 | Molybdenum | Mo | 96·0 |
| Barium | Ba | 137·36 | Nickel | Ni | 58·69 |
| Bismuth | Bi | 209·00 | Nitrogen | N | 14·008 |
| Boron | B | 10·82 | Osmium | Os | 191·5 |
| Bromine | Br | 79·916 | Oxygen | O | 16·000 |
| Cadmium | Cd | 112·41 | Phosphorus | P | 31·02 |
| Calcium | Ca | 40·08 | Platinum | Pt | 195·23 |
| Carbon | C | 12·01 | Potassium | K | 39·096 |
| Chlorine | Cl | 35·457 | Radium | Ra | 225·97 |
| Chromium | Cr | 52·01 | Silicon | Si | 28·06 |
| Cobalt | Co | 58·94 | Silver | Ag | 107·880 |
| Copper | Cu | 63·57 | Sodium | Na | 22·997 |
| Fluorine | F | 19·00 | Strontium | Sr | 87·63 |
| Gold | Au | 197·2 | Sulphur | S | 32·06 |
| Hydrogen | H | 1·0078 | Tin | Sn | 118·70 |
| Iodine | I | 126·92 | Titanium | Ti | 47·90 |
| Iron | Fe | 55·84 | Tungsten | W | 184·0 |
| Lead | Pb | 207·21 | Uranium | U | 238·07 |
| Lithium | Li | 6·94 | Zinc | Zn | 65·38 |
| Magnesium | Mg | 24·32 | | | |

# APPENDIX 3

LOGARITHMS

| | 0 | 1 | 2 | 3 | 4 | 5 | 6 | 7 | 8 | 9 | 1 | 2 | 3 | 4 | 5 | 6 | 7 | 8 | 9 |
|---|---|---|---|---|---|---|---|---|---|---|---|---|---|---|---|---|---|---|---|
| 10 | 0000 | 0043 | 0086 | 0128 | 0170 | 0212 | 0253 | 0294 | 0334 | 0374 | 4 | 8 | 12 | 17 | 21 | 25 | 29 | 33 | 37 |
| 11 | 0414 | 0453 | 0492 | 0531 | 0569 | 0607 | 0645 | 0682 | 0719 | 0755 | 4 | 8 | 11 | 15 | 19 | 23 | 26 | 30 | 34 |
| 12 | 0792 | 0828 | 0864 | 0899 | 0934 | 0969 | 1004 | 1038 | 1072 | 1106 | 3 | 7 | 10 | 14 | 17 | 21 | 24 | 28 | 31 |
| 13 | 1139 | 1173 | 1206 | 1239 | 1271 | 1303 | 1335 | 1367 | 1399 | 1430 | 3 | 6 | 10 | 13 | 16 | 19 | 23 | 26 | 29 |
| 14 | 1461 | 1492 | 1523 | 1553 | 1584 | 1614 | 1644 | 1673 | 1703 | 1732 | 3 | 6 | 9 | 12 | 15 | 18 | 21 | 24 | 27 |
| 15 | 1761 | 1790 | 1818 | 1847 | 1875 | 1903 | 1931 | 1959 | 1987 | 2014 | 3 | 6 | 8 | 11 | 14 | 17 | 20 | 22 | 25 |
| 16 | 2041 | 2068 | 2095 | 2122 | 2148 | 2175 | 2201 | 2227 | 2253 | 2279 | 3 | 5 | 8 | 11 | 13 | 16 | 18 | 21 | 24 |
| 17 | 2304 | 2330 | 2355 | 2380 | 2405 | 2430 | 2455 | 2480 | 2504 | 2529 | 2 | 5 | 7 | 10 | 12 | 15 | 17 | 20 | 22 |
| 18 | 2553 | 2577 | 2601 | 2625 | 2648 | 2672 | 2695 | 2718 | 2742 | 2765 | 2 | 5 | 7 | 9 | 12 | 14 | 16 | 19 | 21 |
| 19 | 2788 | 2810 | 2833 | 2856 | 2878 | 2900 | 2923 | 2945 | 2967 | 2989 | 2 | 4 | 7 | 9 | 11 | 13 | 16 | 18 | 20 |
| 20 | 3010 | 3032 | 3054 | 3075 | 3096 | 3118 | 3139 | 3160 | 3181 | 3201 | 2 | 4 | 6 | 8 | 11 | 13 | 15 | 17 | 19 |
| 21 | 3222 | 3243 | 3263 | 3284 | 3304 | 3324 | 3345 | 3365 | 3385 | 3404 | 2 | 4 | 6 | 8 | 10 | 12 | 14 | 16 | 18 |
| 22 | 3424 | 3444 | 3464 | 3483 | 3502 | 3522 | 3541 | 3560 | 3579 | 3598 | 2 | 4 | 6 | 8 | 10 | 12 | 14 | 15 | 17 |
| 23 | 3617 | 3636 | 3655 | 3674 | 3692 | 3711 | 3729 | 3747 | 3766 | 3784 | 2 | 4 | 6 | 7 | 9 | 11 | 13 | 15 | 17 |
| 24 | 3802 | 3820 | 3838 | 3856 | 3874 | 3892 | 3909 | 3927 | 3945 | 3962 | 2 | 4 | 5 | 7 | 9 | 11 | 12 | 14 | 16 |
| 25 | 3979 | 3997 | 4014 | 4031 | 4048 | 4065 | 4082 | 4099 | 4116 | 4133 | 2 | 3 | 5 | 7 | 9 | 10 | 12 | 14 | 15 |
| 26 | 4150 | 4166 | 4183 | 4200 | 4216 | 4232 | 4249 | 4265 | 4281 | 4298 | 2 | 3 | 5 | 7 | 8 | 10 | 11 | 13 | 15 |
| 27 | 4314 | 4330 | 4346 | 4362 | 4378 | 4393 | 4409 | 4425 | 4440 | 4456 | 2 | 3 | 5 | 6 | 8 | 9 | 11 | 13 | 14 |
| 28 | 4472 | 4487 | 4502 | 4518 | 4533 | 4548 | 4564 | 4579 | 4594 | 4609 | 2 | 3 | 5 | 6 | 8 | 9 | 11 | 12 | 14 |
| 29 | 4624 | 4639 | 4654 | 4669 | 4683 | 4698 | 4713 | 4728 | 4742 | 4757 | 1 | 3 | 4 | 6 | 7 | 9 | 10 | 12 | 13 |
| 30 | 4771 | 4786 | 4800 | 4814 | 4829 | 4843 | 4857 | 4871 | 4886 | 4900 | 1 | 3 | 4 | 6 | 7 | 9 | 10 | 11 | 13 |
| 31 | 4914 | 4928 | 4942 | 4955 | 4969 | 4983 | 4997 | 5011 | 5024 | 5038 | 1 | 3 | 4 | 6 | 7 | 8 | 10 | 11 | 12 |
| 32 | 5051 | 5065 | 5079 | 5092 | 5105 | 5119 | 5132 | 5145 | 5159 | 5172 | 1 | 3 | 4 | 5 | 7 | 8 | 9 | 11 | 12 |
| 33 | 5185 | 5198 | 5211 | 5224 | 5237 | 5250 | 5263 | 5276 | 5289 | 5302 | 1 | 3 | 4 | 5 | 6 | 8 | 9 | 10 | 12 |
| 34 | 5315 | 5328 | 5340 | 5353 | 5366 | 5378 | 5391 | 5403 | 5416 | 5428 | 1 | 3 | 4 | 5 | 6 | 8 | 9 | 10 | 11 |
| 35 | 5441 | 5453 | 5465 | 5478 | 5490 | 5502 | 5514 | 5527 | 5539 | 5551 | 1 | 2 | 4 | 5 | 6 | 7 | 9 | 10 | 11 |
| 36 | 5563 | 5575 | 5587 | 5599 | 5611 | 5623 | 5635 | 5647 | 5658 | 5670 | 1 | 2 | 4 | 5 | 6 | 7 | 8 | 10 | 11 |
| 37 | 5682 | 5694 | 5705 | 5717 | 5729 | 5740 | 5752 | 5763 | 5775 | 5786 | 1 | 2 | 3 | 5 | 6 | 7 | 8 | 9 | 10 |
| 38 | 5798 | 5809 | 5821 | 5832 | 5843 | 5855 | 5866 | 5877 | 5888 | 5899 | 1 | 2 | 3 | 5 | 6 | 7 | 8 | 9 | 10 |
| 39 | 5911 | 5922 | 5933 | 5944 | 5955 | 5966 | 5977 | 5988 | 5999 | 6010 | 1 | 2 | 3 | 4 | 5 | 7 | 8 | 9 | 10 |
| 40 | 6021 | 6031 | 6042 | 6053 | 6064 | 6075 | 6085 | 6096 | 6107 | 6117 | 1 | 2 | 3 | 4 | 5 | 6 | 8 | 9 | 10 |
| 41 | 6128 | 6138 | 6149 | 6160 | 6170 | 6180 | 6191 | 6201 | 6212 | 6222 | 1 | 2 | 3 | 4 | 5 | 6 | 7 | 8 | 9 |
| 42 | 6232 | 6243 | 6253 | 6263 | 6274 | 6284 | 6294 | 6304 | 6314 | 6325 | 1 | 2 | 3 | 4 | 5 | 6 | 7 | 8 | 9 |
| 43 | 6335 | 6345 | 6355 | 6365 | 6375 | 6385 | 6395 | 6405 | 6415 | 6425 | 1 | 2 | 3 | 4 | 5 | 6 | 7 | 8 | 9 |
| 44 | 6435 | 6444 | 6454 | 6464 | 6474 | 6484 | 6493 | 6503 | 6513 | 6522 | 1 | 2 | 3 | 4 | 5 | 6 | 7 | 8 | 9 |
| 45 | 6532 | 6542 | 6551 | 6561 | 6571 | 6580 | 6590 | 6599 | 6609 | 6618 | 1 | 2 | 3 | 4 | 5 | 6 | 7 | 8 | 9 |
| 46 | 6628 | 6637 | 6646 | 6656 | 6665 | 6675 | 6684 | 6693 | 6702 | 6712 | 1 | 2 | 3 | 4 | 5 | 6 | 7 | 7 | 8 |
| 47 | 6721 | 6730 | 6739 | 6749 | 6758 | 6767 | 6776 | 6785 | 6794 | 6803 | 1 | 2 | 3 | 4 | 5 | 5 | 6 | 7 | 8 |
| 48 | 6812 | 6821 | 6830 | 6839 | 6848 | 6857 | 6866 | 6875 | 6884 | 6893 | 1 | 2 | 3 | 4 | 4 | 5 | 6 | 7 | 8 |
| 49 | 6902 | 6911 | 6920 | 6928 | 6937 | 6946 | 6955 | 6964 | 6972 | 6981 | 1 | 2 | 3 | 4 | 4 | 5 | 6 | 7 | 8 |
| 50 | 6990 | 6998 | 7007 | 7016 | 7024 | 7033 | 7042 | 7050 | 7059 | 7067 | 1 | 2 | 3 | 3 | 4 | 5 | 6 | 7 | 8 |
| 51 | 7076 | 7084 | 7093 | 7101 | 7110 | 7118 | 7126 | 7135 | 7143 | 7152 | 1 | 2 | 3 | 3 | 4 | 5 | 6 | 7 | 8 |
| 52 | 7160 | 7168 | 7177 | 7185 | 7193 | 7202 | 7210 | 7218 | 7226 | 7235 | 1 | 2 | 2 | 3 | 4 | 5 | 6 | 7 | 7 |
| 53 | 7243 | 7251 | 7259 | 7267 | 7275 | 7284 | 7292 | 7300 | 7308 | 7316 | 1 | 2 | 2 | 3 | 4 | 5 | 6 | 6 | 7 |
| 54 | 7324 | 7332 | 7340 | 7348 | 7356 | 7364 | 7372 | 7380 | 7388 | 7396 | 1 | 2 | 2 | 3 | 4 | 5 | 6 | 6 | 7 |

## LOGARITHMS

|    | 0 | 1 | 2 | 3 | 4 | 5 | 6 | 7 | 8 | 9 | 1 | 2 | 3 | 4 | 5 | 6 | 7 | 8 | 9 |
|----|---|---|---|---|---|---|---|---|---|---|---|---|---|---|---|---|---|---|---|
| 55 | 7404 | 7412 | 7419 | 7427 | 7435 | 7443 | 7451 | 7459 | 7466 | 7474 | 1 | 2 | 2 | 3 | 4 | 5 | 5 | 6 | 7 |
| 56 | 7482 | 7490 | 7497 | 7505 | 7513 | 7520 | 7528 | 7536 | 7543 | 7551 | 1 | 2 | 2 | 3 | 4 | 5 | 5 | 6 | 7 |
| 57 | 7559 | 7566 | 7574 | 7582 | 7589 | 7597 | 7604 | 7612 | 7619 | 7627 | 1 | 2 | 2 | 3 | 4 | 5 | 5 | 6 | 7 |
| 58 | 7634 | 7642 | 7649 | 7657 | 7664 | 7672 | 7679 | 7686 | 7694 | 7701 | 1 | 1 | 2 | 3 | 4 | 4 | 5 | 6 | 7 |
| 59 | 7709 | 7716 | 7723 | 7731 | 7738 | 7745 | 7752 | 7760 | 7767 | 7774 | 1 | 1 | 2 | 3 | 4 | 4 | 5 | 6 | 7 |
| 60 | 7782 | 7789 | 7796 | 7803 | 7810 | 7818 | 7825 | 7832 | 7839 | 7846 | 1 | 1 | 2 | 3 | 4 | 4 | 5 | 6 | 6 |
| 61 | 7853 | 7860 | 7868 | 7875 | 7882 | 7889 | 7896 | 7903 | 7910 | 7917 | 1 | 1 | 2 | 3 | 4 | 4 | 5 | 6 | 6 |
| 62 | 7924 | 7931 | 7938 | 7945 | 7952 | 7959 | 7966 | 7973 | 7980 | 7987 | 1 | 1 | 2 | 3 | 3 | 4 | 5 | 6 | 6 |
| 63 | 7993 | 8000 | 8007 | 8014 | 8021 | 8028 | 8035 | 8041 | 8048 | 8055 | 1 | 1 | 2 | 3 | 3 | 4 | 5 | 5 | 6 |
| 64 | 8062 | 8069 | 8075 | 8082 | 8089 | 8096 | 8102 | 8109 | 8116 | 8122 | 1 | 1 | 2 | 3 | 3 | 4 | 5 | 5 | 6 |
| 65 | 8129 | 8136 | 8142 | 8149 | 8156 | 8162 | 8169 | 8176 | 8182 | 8189 | 1 | 1 | 2 | 3 | 3 | 4 | 5 | 5 | 6 |
| 66 | 8195 | 8202 | 8209 | 8215 | 8222 | 8228 | 8235 | 8241 | 8248 | 8254 | 1 | 1 | 2 | 3 | 3 | 4 | 5 | 5 | 6 |
| 67 | 8261 | 8267 | 8274 | 8280 | 8287 | 8293 | 8299 | 8306 | 8312 | 8319 | 1 | 1 | 2 | 3 | 3 | 4 | 5 | 5 | 6 |
| 68 | 8325 | 8331 | 8338 | 8344 | 8351 | 8357 | 8363 | 8370 | 8376 | 8382 | 1 | 1 | 2 | 3 | 3 | 4 | 4 | 5 | 6 |
| 69 | 8388 | 8395 | 8401 | 8407 | 8414 | 8420 | 8426 | 8432 | 8439 | 8445 | 1 | 1 | 2 | 2 | 3 | 4 | 4 | 5 | 6 |
| 70 | 8451 | 8457 | 8463 | 8470 | 8476 | 8482 | 8488 | 8494 | 8500 | 8506 | 1 | 1 | 2 | 2 | 3 | 4 | 4 | 5 | 6 |
| 71 | 8513 | 8519 | 8525 | 8531 | 8537 | 8543 | 8549 | 8555 | 8561 | 8567 | 1 | 1 | 2 | 2 | 3 | 4 | 4 | 5 | 5 |
| 72 | 8573 | 8579 | 8585 | 8591 | 8597 | 8603 | 8609 | 8615 | 8621 | 8627 | 1 | 1 | 2 | 2 | 3 | 4 | 4 | 5 | 5 |
| 73 | 8633 | 8639 | 8645 | 8651 | 8657 | 8663 | 8669 | 8675 | 8681 | 8686 | 1 | 1 | 2 | 2 | 3 | 4 | 4 | 5 | 5 |
| 74 | 8692 | 8698 | 8704 | 8710 | 8716 | 8722 | 8727 | 8733 | 8739 | 8745 | 1 | 1 | 2 | 2 | 3 | 4 | 4 | 5 | 5 |
| 75 | 8751 | 8756 | 8762 | 8768 | 8774 | 8779 | 8785 | 8791 | 8797 | 8802 | 1 | 1 | 2 | 2 | 3 | 3 | 4 | 5 | 5 |
| 76 | 8808 | 8814 | 8820 | 8825 | 8831 | 8837 | 8842 | 8848 | 8854 | 8859 | 1 | 1 | 2 | 2 | 3 | 3 | 4 | 5 | 5 |
| 77 | 8865 | 8871 | 8876 | 8882 | 8887 | 8893 | 8899 | 8904 | 8910 | 8915 | 1 | 1 | 2 | 2 | 3 | 3 | 4 | 4 | 5 |
| 78 | 8921 | 8927 | 8932 | 8938 | 8943 | 8949 | 8954 | 8960 | 8965 | 8971 | 1 | 1 | 2 | 2 | 3 | 3 | 4 | 4 | 5 |
| 79 | 8976 | 8982 | 8987 | 8993 | 8998 | 9004 | 9009 | 9015 | 9020 | 9025 | 1 | 1 | 2 | 2 | 3 | 3 | 4 | 4 | 5 |
| 80 | 9031 | 9036 | 9042 | 9047 | 9053 | 9058 | 9063 | 9069 | 9074 | 9079 | 1 | 1 | 2 | 2 | 3 | 3 | 4 | 4 | 5 |
| 81 | 9085 | 9090 | 9096 | 9101 | 9106 | 9112 | 9117 | 9122 | 9128 | 9133 | 1 | 1 | 2 | 2 | 3 | 3 | 4 | 4 | 5 |
| 82 | 9138 | 9143 | 9149 | 9154 | 9159 | 9165 | 9170 | 9175 | 9180 | 9186 | 1 | 1 | 2 | 2 | 3 | 3 | 4 | 4 | 5 |
| 83 | 9191 | 9196 | 9201 | 9206 | 9212 | 9217 | 9222 | 9227 | 9232 | 9238 | 1 | 1 | 2 | 2 | 3 | 3 | 4 | 4 | 5 |
| 84 | 9243 | 9248 | 9253 | 9258 | 9263 | 9269 | 9274 | 9279 | 9284 | 9289 | 1 | 1 | 2 | 2 | 3 | 3 | 4 | 4 | 5 |
| 85 | 9294 | 9299 | 9304 | 9309 | 9315 | 9320 | 9325 | 9330 | 9335 | 9340 | 1 | 1 | 2 | 2 | 3 | 3 | 4 | 4 | 5 |
| 86 | 9345 | 9350 | 9355 | 9360 | 9335 | 9370 | 9375 | 9380 | 9385 | 9390 | 1 | 1 | 2 | 2 | 3 | 3 | 4 | 4 | 5 |
| 87 | 9395 | 9400 | 9405 | 9410 | 9415 | 9420 | 9425 | 9430 | 9435 | 9440 | 0 | 1 | 1 | 2 | 2 | 3 | 3 | 4 | 4 |
| 88 | 9445 | 9450 | 9455 | 9460 | 9465 | 9469 | 9474 | 9479 | 9484 | 9489 | 0 | 1 | 1 | 2 | 2 | 3 | 3 | 4 | 4 |
| 89 | 9494 | 9499 | 9504 | 9509 | 9513 | 9518 | 9523 | 9528 | 9533 | 9538 | 0 | 1 | 1 | 2 | 2 | 3 | 3 | 4 | 4 |
| 90 | 9542 | 9547 | 9552 | 9557 | 9562 | 9566 | 9571 | 9576 | 9581 | 9586 | 0 | 1 | 1 | 2 | 2 | 3 | 3 | 4 | 4 |
| 91 | 9590 | 9595 | 9600 | 9605 | 9609 | 9614 | 9619 | 9624 | 9628 | 9633 | 0 | 1 | 1 | 2 | 2 | 3 | 3 | 4 | 4 |
| 92 | 9638 | 9643 | 9647 | 9652 | 9657 | 9661 | 9666 | 9671 | 9675 | 9680 | 0 | 1 | 1 | 2 | 2 | 3 | 3 | 4 | 4 |
| 93 | 9685 | 9689 | 9694 | 9699 | 9703 | 9708 | 9713 | 9717 | 9722 | 9727 | 0 | 1 | 1 | 2 | 2 | 3 | 3 | 4 | 4 |
| 94 | 9731 | 9736 | 9741 | 9745 | 9750 | 9754 | 9759 | 9763 | 9768 | 9773 | 0 | 1 | 1 | 2 | 2 | 3 | 3 | 4 | 4 |
| 95 | 9777 | 9782 | 9786 | 9791 | 9795 | 9800 | 9805 | 9809 | 9814 | 9818 | 0 | 1 | 1 | 2 | 2 | 3 | 3 | 4 | 4 |
| 96 | 9823 | 9827 | 9832 | 9836 | 9841 | 9845 | 9850 | 9854 | 9858 | 9863 | 0 | 1 | 1 | 2 | 2 | 3 | 3 | 4 | 4 |
| 97 | 9868 | 9872 | 9877 | 9881 | 9886 | 9890 | 9894 | 9899 | 9903 | 9908 | 0 | 1 | 1 | 2 | 2 | 3 | 3 | 4 | 4 |
| 98 | 9912 | 9917 | 9921 | 9926 | 9930 | 9934 | 9939 | 9943 | 9948 | 9952 | 0 | 1 | 1 | 2 | 2 | 3 | 3 | 4 | 4 |
| 99 | 9956 | 9961 | 9965 | 9969 | 9974 | 9978 | 9983 | 9987 | 9991 | 9916 | 0 | 1 | 1 | 2 | 2 | 3 | 3 | 3 | 4 |

ANTILOGARITHMS

| | 0 | 1 | 2 | 3 | 4 | 5 | 6 | 7 | 8 | 9 | 1 | 2 | 3 | 4 | 5 | 6 | 7 | 8 | 9 |
|---|---|---|---|---|---|---|---|---|---|---|---|---|---|---|---|---|---|---|---|
| ·00 | 1000 | 1002 | 1005 | 1007 | 1009 | 1012 | 1014 | 1016 | 1019 | 1021 | 0 | 0 | 1 | 1 | 1 | 1 | 2 | 2 | 2 |
| ·01 | 1023 | 1026 | 1028 | 1030 | 1033 | 1035 | 1038 | 1040 | 1042 | 1045 | 0 | 0 | 1 | 1 | 1 | 1 | 2 | 2 | 2 |
| ·02 | 1047 | 1050 | 1052 | 1054 | 1057 | 1059 | 1062 | 1064 | 1067 | 1069 | 0 | 0 | 1 | 1 | 1 | 1 | 2 | 2 | 2 |
| ·03 | 1072 | 1074 | 1076 | 1079 | 1081 | 1084 | 1086 | 1089 | 1091 | 1094 | 0 | 0 | 1 | 1 | 1 | 1 | 2 | 2 | 2 |
| ·04 | 1095 | 1099 | 1102 | 1104 | 1107 | 1109 | 1112 | 1114 | 1117 | 1119 | 0 | 1 | 1 | 1 | 1 | 2 | 2 | 2 | 2 |
| ·05 | 1122 | 1125 | 1127 | 1130 | 1132 | 1135 | 1138 | 1140 | 1143 | 1146 | 0 | 1 | 1 | 1 | 1 | 2 | 2 | 2 | 2 |
| ·06 | 1148 | 1151 | 1153 | 1156 | 1159 | 1161 | 1164 | 1167 | 1169 | 1172 | 0 | 1 | 1 | 1 | 1 | 2 | 2 | 2 | 2 |
| ·07 | 1175 | 1178 | 1180 | 1183 | 1186 | 1189 | 1191 | 1194 | 1197 | 1199 | 0 | 1 | 1 | 1 | 1 | 2 | 2 | 2 | 2 |
| ·08 | 1202 | 1205 | 1208 | 1211 | 1213 | 1216 | 1219 | 1222 | 1225 | 1227 | 0 | 1 | 1 | 1 | 1 | 2 | 2 | 2 | 3 |
| ·09 | 1230 | 1233 | 1236 | 1239 | 1242 | 1245 | 1247 | 1250 | 1253 | 1256 | 0 | 1 | 1 | 1 | 1 | 2 | 2 | 2 | 3 |
| ·10 | 1259 | 1262 | 1265 | 1268 | 1271 | 1274 | 1276 | 1279 | 1282 | 1285 | 0 | 1 | 1 | 1 | 1 | 2 | 2 | 2 | 3 |
| ·11 | 1288 | 1291 | 1294 | 1297 | 1300 | 1303 | 1306 | 1309 | 1312 | 1315 | 0 | 1 | 1 | 1 | 2 | 2 | 2 | 2 | 3 |
| ·12 | 1318 | 1321 | 1324 | 1327 | 1330 | 1334 | 1337 | 1340 | 1343 | 1346 | 0 | 1 | 1 | 1 | 2 | 2 | 2 | 2 | 3 |
| ·13 | 1349 | 1352 | 1355 | 1358 | 1361 | 1365 | 1368 | 1371 | 1374 | 1377 | 0 | 1 | 1 | 1 | 2 | 2 | 2 | 3 | 3 |
| ·14 | 1380 | 1384 | 1387 | 1390 | 1393 | 1396 | 1400 | 1403 | 1406 | 1409 | 0 | 1 | 1 | 1 | 2 | 2 | 2 | 3 | 3 |
| ·15 | 1413 | 1416 | 1419 | 1422 | 1426 | 1429 | 1432 | 1435 | 1439 | 1442 | 0 | 1 | 1 | 1 | 2 | 2 | 2 | 3 | 3 |
| ·16 | 1445 | 1449 | 1452 | 1455 | 1459 | 1462 | 1466 | 1469 | 1472 | 1476 | 0 | 1 | 1 | 1 | 2 | 2 | 2 | 3 | 3 |
| ·17 | 1479 | 1483 | 1486 | 1489 | 1493 | 1496 | 1500 | 1503 | 1507 | 1510 | 0 | 1 | 1 | 1 | 2 | 2 | 2 | 3 | 3 |
| ·18 | 1514 | 1517 | 1521 | 1524 | 1528 | 1531 | 1535 | 1538 | 1542 | 1545 | 0 | 1 | 1 | 1 | 2 | 2 | 2 | 3 | 3 |
| ·19 | 1549 | 1552 | 1556 | 1560 | 1563 | 1567 | 1570 | 1574 | 1578 | 1581 | 0 | 1 | 1 | 1 | 2 | 2 | 3 | 3 | 3 |
| ·20 | 1585 | 1589 | 1592 | 1596 | 1600 | 1603 | 1607 | 1611 | 1614 | 1618 | 0 | 1 | 1 | 1 | 2 | 2 | 3 | 3 | 3 |
| ·21 | 1622 | 1626 | 1629 | 1633 | 1637 | 1641 | 1644 | 1648 | 1652 | 1656 | 0 | 1 | 1 | 2 | 2 | 2 | 3 | 3 | 3 |
| ·22 | 1660 | 1663 | 1667 | 1671 | 1675 | 1679 | 1683 | 1687 | 1690 | 1694 | 0 | 1 | 1 | 2 | 2 | 2 | 3 | 3 | 3 |
| ·23 | 1698 | 1702 | 1706 | 1710 | 1714 | 1718 | 1722 | 1726 | 1730 | 1734 | 0 | 1 | 1 | 2 | 2 | 2 | 3 | 3 | 4 |
| ·24 | 1738 | 1742 | 1746 | 1750 | 1754 | 1758 | 1762 | 1766 | 1770 | 1774 | 0 | 1 | 1 | 2 | 2 | 2 | 3 | 3 | 4 |
| ·25 | 1778 | 1782 | 1786 | 1791 | 1795 | 1799 | 1803 | 1807 | 1811 | 1816 | 0 | 1 | 1 | 2 | 2 | 2 | 3 | 3 | 4 |
| ·26 | 1820 | 1824 | 1828 | 1832 | 1837 | 1841 | 1845 | 1848 | 1854 | 1858 | 0 | 1 | 1 | 2 | 2 | 3 | 3 | 3 | 4 |
| ·27 | 1862 | 1866 | 1871 | 1875 | 1879 | 1884 | 1888 | 1892 | 1897 | 1901 | 0 | 1 | 1 | 2 | 2 | 3 | 3 | 3 | 4 |
| ·28 | 1905 | 1910 | 1914 | 1919 | 1923 | 1928 | 1932 | 1936 | 1941 | 1945 | 0 | 1 | 1 | 2 | 2 | 3 | 3 | 4 | 4 |
| ·29 | 1950 | 1954 | 1959 | 1963 | 1968 | 1972 | 1977 | 1982 | 1986 | 1991 | 0 | 1 | 1 | 2 | 2 | 3 | 3 | 4 | 4 |
| ·30 | 1995 | 2000 | 2004 | 2009 | 2014 | 2018 | 2023 | 2028 | 2032 | 2037 | 0 | 1 | 1 | 2 | 2 | 3 | 3 | 4 | 4 |
| ·31 | 2042 | 2046 | 2051 | 2056 | 2061 | 2065 | 2070 | 2075 | 2080 | 2084 | 0 | 1 | 1 | 2 | 2 | 3 | 3 | 4 | 4 |
| ·32 | 2089 | 2094 | 2099 | 2104 | 2109 | 2113 | 2118 | 2123 | 2128 | 2133 | 0 | 1 | 1 | 2 | 2 | 3 | 3 | 4 | 4 |
| ·33 | 2138 | 2143 | 2148 | 2153 | 2158 | 2163 | 2168 | 2173 | 2178 | 2183 | 0 | 1 | 1 | 2 | 2 | 3 | 3 | 4 | 4 |
| ·34 | 2188 | 2193 | 2198 | 2203 | 2208 | 2213 | 2218 | 2223 | 2228 | 2234 | 1 | 1 | 2 | 2 | 3 | 3 | 4 | 4 | 5 |
| ·35 | 2239 | 2244 | 2249 | 2254 | 2259 | 2265 | 2270 | 2275 | 2280 | 2286 | 1 | 1 | 2 | 2 | 3 | 3 | 4 | 4 | 5 |
| ·36 | 2291 | 2296 | 2301 | 2307 | 2312 | 2317 | 2323 | 2328 | 2333 | 2339 | 1 | 1 | 2 | 2 | 3 | 3 | 4 | 4 | 5 |
| ·37 | 2344 | 2350 | 2355 | 2360 | 2366 | 2371 | 2377 | 2382 | 2388 | 2393 | 1 | 1 | 2 | 2 | 3 | 3 | 4 | 4 | 5 |
| ·38 | 2399 | 2404 | 2410 | 2415 | 2421 | 2427 | 2432 | 2438 | 2443 | 2449 | 1 | 1 | 2 | 2 | 3 | 3 | 4 | 4 | 5 |
| ·39 | 2455 | 2460 | 2466 | 2472 | 2477 | 2483 | 2489 | 2495 | 2500 | 2506 | 1 | 1 | 2 | 2 | 3 | 3 | 4 | 5 | 5 |
| ·40 | 2512 | 2518 | 2523 | 2529 | 2535 | 2541 | 2547 | 2553 | 2559 | 2564 | 1 | 1 | 2 | 2 | 3 | 4 | 4 | 5 | 5 |
| ·41 | 2570 | 2576 | 2582 | 2588 | 2594 | 2600 | 2606 | 2612 | 2618 | 2624 | 1 | 1 | 2 | 2 | 3 | 4 | 4 | 5 | 5 |
| ·42 | 2630 | 2636 | 2642 | 2649 | 2655 | 2661 | 2667 | 2673 | 2679 | 2685 | 1 | 1 | 2 | 2 | 3 | 4 | 4 | 5 | 6 |
| ·43 | 2692 | 2698 | 2704 | 2710 | 2716 | 2723 | 2729 | 2735 | 2742 | 2748 | 1 | 1 | 2 | 3 | 3 | 4 | 4 | 5 | 6 |
| ·44 | 2754 | 2761 | 2767 | 2773 | 2780 | 2786 | 2793 | 2799 | 2805 | 2812 | 1 | 1 | 2 | 3 | 3 | 4 | 4 | 5 | 6 |
| ·45 | 2818 | 2825 | 2831 | 2838 | 2844 | 2851 | 2858 | 2864 | 2871 | 2877 | 1 | 1 | 2 | 3 | 3 | 4 | 5 | 5 | 6 |
| ·46 | 2884 | 2891 | 2897 | 2904 | 2911 | 2917 | 2924 | 2931 | 2938 | 2944 | 1 | 1 | 2 | 3 | 3 | 4 | 5 | 5 | 6 |
| ·47 | 2951 | 2958 | 2965 | 2972 | 2979 | 2985 | 2992 | 2999 | 3006 | 3013 | 1 | 1 | 2 | 3 | 3 | 4 | 5 | 5 | 6 |
| ·48 | 3020 | 3027 | 3034 | 3041 | 3048 | 3055 | 3062 | 3069 | 3076 | 3083 | 1 | 1 | 2 | 3 | 4 | 4 | 5 | 6 | 6 |
| ·49 | 3090 | 3097 | 3105 | 3112 | 3119 | 3126 | 3133 | 3141 | 3148 | 3155 | 1 | 1 | 2 | 3 | 4 | 4 | 5 | 6 | 6 |

ANTILOGARITHMS

| | 0 | 1 | 2 | 3 | 4 | 5 | 6 | 7 | 8 | 9 | 1 | 2 | 3 | 4 | 5 | 6 | 7 | 8 | 9 |
|---|---|---|---|---|---|---|---|---|---|---|---|---|---|---|---|---|---|---|---|
| ·50 | 3162 | 3170 | 3177 | 3184 | 3192 | 3199 | 3206 | 3214 | 3221 | 3228 | 1 | 1 | 2 | 3 | 4 | 4 | 5 | 6 | 7 |
| ·51 | 3236 | 3243 | 3251 | 3258 | 3266 | 3273 | 3281 | 3289 | 3296 | 3304 | 1 | 2 | 2 | 3 | 4 | 5 | 5 | 6 | 7 |
| ·52 | 3311 | 3319 | 3327 | 3334 | 3342 | 3350 | 3357 | 3365 | 3373 | 3381 | 1 | 2 | 2 | 3 | 4 | 5 | 5 | 6 | 7 |
| ·53 | 3388 | 3396 | 3404 | 3412 | 3420 | 3428 | 3436 | 3443 | 3451 | 3459 | 1 | 2 | 2 | 3 | 4 | 5 | 6 | 6 | 7 |
| ·54 | 3467 | 3475 | 3483 | 3491 | 3499 | 3508 | 3516 | 3524 | 3532 | 3540 | 1 | 2 | 2 | 3 | 4 | 5 | 6 | 6 | 7 |
| ·55 | 3548 | 3556 | 3565 | 3573 | 3581 | 3589 | 3597 | 3606 | 3614 | 3622 | 1 | 2 | 2 | 3 | 4 | 5 | 6 | 7 | 7 |
| ·56 | 3631 | 3639 | 3648 | 3656 | 3664 | 3673 | 3681 | 3690 | 3698 | 3707 | 1 | 2 | 3 | 3 | 4 | 5 | 6 | 7 | 8 |
| ·57 | 3715 | 3724 | 3733 | 3741 | 3750 | 3758 | 3767 | 3776 | 3784 | 3793 | 1 | 2 | 3 | 3 | 4 | 5 | 6 | 7 | 8 |
| ·58 | 3802 | 3811 | 3819 | 3828 | 3837 | 3846 | 3855 | 3864 | 3873 | 3882 | 1 | 2 | 3 | 4 | 4 | 5 | 6 | 7 | 8 |
| ·59 | 3890 | 3899 | 3908 | 3917 | 3926 | 3936 | 3945 | 3954 | 3963 | 3972 | 1 | 2 | 3 | 4 | 5 | 5 | 6 | 7 | 8 |
| ·60 | 3981 | 3990 | 3999 | 4009 | 4018 | 4027 | 4036 | 4046 | 4055 | 4064 | 1 | 2 | 3 | 4 | 5 | 6 | 6 | 7 | 8 |
| ·61 | 4074 | 4083 | 4093 | 4102 | 4111 | 4121 | 4130 | 4140 | 4150 | 4159 | 1 | 2 | 3 | 4 | 5 | 6 | 7 | 8 | 9 |
| ·62 | 4169 | 4178 | 4188 | 4198 | 4207 | 4217 | 4227 | 4236 | 4246 | 4256 | 1 | 2 | 3 | 4 | 5 | 6 | 7 | 8 | 9 |
| ·63 | 4266 | 4276 | 4285 | 4295 | 4305 | 4315 | 4325 | 4335 | 4345 | 4355 | 1 | 2 | 3 | 4 | 5 | 6 | 7 | 8 | 9 |
| ·64 | 4365 | 4375 | 4385 | 4395 | 4406 | 4416 | 4426 | 4436 | 4446 | 4457 | 1 | 2 | 3 | 4 | 5 | 6 | 7 | 8 | 9 |
| ·65 | 4467 | 4477 | 4487 | 4498 | 4508 | 4519 | 4529 | 4539 | 4550 | 4560 | 1 | 2 | 3 | 4 | 5 | 6 | 7 | 8 | 9 |
| ·66 | 4571 | 4581 | 4592 | 4603 | 4613 | 4624 | 4634 | 4645 | 4656 | 4667 | 1 | 2 | 3 | 4 | 5 | 6 | 7 | 9 | 10 |
| ·67 | 4677 | 4688 | 4699 | 4710 | 4721 | 4732 | 4742 | 4753 | 4764 | 4775 | 1 | 2 | 3 | 4 | 5 | 7 | 8 | 9 | 10 |
| ·68 | 4786 | 4797 | 4808 | 4819 | 4831 | 4842 | 4853 | 4864 | 4875 | 4887 | 1 | 2 | 3 | 4 | 6 | 7 | 8 | 9 | 10 |
| ·69 | 4898 | 4909 | 4920 | 4932 | 4943 | 4955 | 4966 | 4977 | 4989 | 5000 | 1 | 2 | 3 | 5 | 6 | 7 | 8 | 9 | 10 |
| ·70 | 5012 | 5023 | 5035 | 5047 | 5058 | 5070 | 5082 | 5093 | 5105 | 5117 | 1 | 2 | 4 | 5 | 6 | 7 | 8 | 9 | 11 |
| ·71 | 5129 | 5140 | 5152 | 5164 | 5176 | 5188 | 5200 | 5212 | 5224 | 5236 | 1 | 2 | 4 | 5 | 6 | 7 | 8 | 10 | 11 |
| ·72 | 5248 | 5260 | 5272 | 5284 | 5297 | 5309 | 5321 | 5333 | 5346 | 5358 | 1 | 2 | 4 | 5 | 6 | 7 | 9 | 10 | 11 |
| ·73 | 5370 | 5383 | 5395 | 5408 | 5420 | 5433 | 5445 | 5458 | 5470 | 5483 | 1 | 3 | 4 | 5 | 6 | 8 | 9 | 10 | 11 |
| ·74 | 5495 | 5508 | 5521 | 5534 | 5546 | 5559 | 5572 | 5585 | 5598 | 5610 | 1 | 3 | 4 | 5 | 6 | 8 | 9 | 10 | 12 |
| ·75 | 5623 | 5636 | 5649 | 5662 | 5675 | 5689 | 5702 | 5715 | 5728 | 5741 | 1 | 3 | 4 | 5 | 7 | 8 | 9 | 10 | 12 |
| ·76 | 5754 | 5768 | 5781 | 5794 | 5808 | 5821 | 5834 | 5848 | 5861 | 5875 | 1 | 3 | 4 | 5 | 7 | 8 | 9 | 11 | 12 |
| ·77 | 5888 | 5902 | 5916 | 5929 | 5943 | 5957 | 5970 | 5984 | 5998 | 6012 | 1 | 3 | 4 | 5 | 7 | 8 | 10 | 11 | 12 |
| ·78 | 6026 | 6039 | 6053 | 6067 | 6081 | 6095 | 6109 | 6124 | 6138 | 6152 | 1 | 3 | 4 | 6 | 7 | 8 | 10 | 11 | 13 |
| ·79 | 6166 | 6180 | 6194 | 6209 | 6223 | 6237 | 6252 | 6266 | 6281 | 6295 | 1 | 3 | 4 | 6 | 7 | 9 | 10 | 11 | 13 |
| ·80 | 6310 | 6324 | 6339 | 6353 | 6368 | 6383 | 6397 | 6412 | 6427 | 6442 | 1 | 3 | 4 | 6 | 7 | 9 | 10 | 12 | 13 |
| ·81 | 6457 | 6471 | 6486 | 6531 | 6516 | 6531 | 6546 | 6561 | 6577 | 6592 | 2 | 3 | 5 | 6 | 8 | 9 | 11 | 12 | 14 |
| ·82 | 6607 | 6622 | 6637 | 6653 | 6668 | 6683 | 6699 | 6714 | 6730 | 6745 | 2 | 3 | 5 | 6 | 8 | 9 | 11 | 12 | 14 |
| ·83 | 6761 | 6776 | 6792 | 6808 | 6823 | 6839 | 6855 | 6871 | 6887 | 6902 | 2 | 3 | 5 | 6 | 8 | 9 | 11 | 13 | 14 |
| ·84 | 6918 | 6934 | 6950 | 6966 | 6982 | 6998 | 7015 | 7031 | 7047 | 7063 | 2 | 3 | 5 | 6 | 8 | 10 | 11 | 13 | 15 |
| ·85 | 7079 | 7096 | 7112 | 7129 | 7145 | 7161 | 7178 | 7194 | 7211 | 7228 | 2 | 3 | 5 | 7 | 8 | 10 | 12 | 13 | 15 |
| ·86 | 7244 | 7261 | 7278 | 7295 | 7311 | 7328 | 7345 | 7362 | 7379 | 7396 | 2 | 3 | 5 | 7 | 8 | 10 | 12 | 13 | 15 |
| ·87 | 7413 | 7430 | 7447 | 7464 | 7482 | 7499 | 7516 | 7534 | 7551 | 7568 | 2 | 4 | 5 | 7 | 9 | 10 | 12 | 14 | 16 |
| ·88 | 7586 | 7603 | 7621 | 7638 | 7656 | 7674 | 7691 | 7709 | 7727 | 7745 | 2 | 4 | 5 | 7 | 9 | 11 | 12 | 14 | 16 |
| ·89 | 7762 | 7780 | 7798 | 7816 | 7834 | 7852 | 7870 | 7889 | 7907 | 7925 | 2 | 4 | 5 | 7 | 9 | 11 | 13 | 14 | 16 |
| ·90 | 7943 | 7962 | 7980 | 7998 | 8017 | 8035 | 8054 | 8072 | 8091 | 8110 | 2 | 4 | 6 | 7 | 9 | 11 | 13 | 15 | 17 |
| ·91 | 8128 | 8147 | 8166 | 8185 | 8204 | 8222 | 8241 | 8260 | 8279 | 8299 | 2 | 4 | 6 | 8 | 9 | 11 | 13 | 15 | 17 |
| ·92 | 8318 | 8337 | 8356 | 8375 | 8395 | 8414 | 8433 | 8453 | 8472 | 8492 | 2 | 4 | 6 | 8 | 10 | 12 | 14 | 15 | 17 |
| ·93 | 8511 | 8531 | 8551 | 8570 | 8590 | 8610 | 8630 | 8650 | 8670 | 8690 | 2 | 4 | 6 | 8 | 10 | 12 | 14 | 16 | 18 |
| ·94 | 8710 | 8730 | 8750 | 8770 | 8790 | 8810 | 8831 | 8851 | 8872 | 8892 | 2 | 4 | 6 | 8 | 10 | 12 | 14 | 16 | 18 |
| ·95 | 8913 | 8933 | 8954 | 8974 | 8995 | 9016 | 9034 | 9057 | 9078 | 9099 | 2 | 4 | 6 | 8 | 10 | 12 | 15 | 17 | 19 |
| ·96 | 9120 | 9141 | 9162 | 9183 | 9204 | 9226 | 9247 | 9268 | 9290 | 9311 | 2 | 4 | 6 | 8 | 11 | 13 | 15 | 17 | 19 |
| ·97 | 9333 | 9354 | 9376 | 9397 | 9419 | 9441 | 9462 | 9484 | 9506 | 9528 | 2 | 4 | 7 | 9 | 11 | 13 | 15 | 17 | 20 |
| ·98 | 9550 | 9572 | 9594 | 9616 | 9638 | 9661 | 9683 | 9705 | 9727 | 9750 | 2 | 4 | 7 | 9 | 11 | 13 | 16 | 18 | 20 |
| ·99 | 9772 | 9795 | 9817 | 9840 | 9863 | 9886 | 9908 | 9931 | 9954 | 9977 | 2 | 5 | 7 | 9 | 11 | 14 | 16 | 18 | 20 |

# INDEX

*Numbers in bold type refer to pages with illustrations*